Small World, Isn't It?

Slope, Derivatives, and Exponential Growth

Teacher's Guide

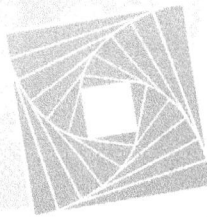

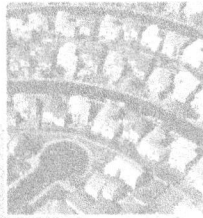

This material is based upon work supported by the National Science Foundation under award numbers ESI-9255262, ESI-0137805, and ESI-0627821. Any opinions, findings, and conclusions or recommendations expressed in this publication are those of the authors and do not necessarily reflect the views of the National Science Foundation.

Key Curriculum
1150 65th Street
Emeryville, California 94608
email: editorial@keypress.com
www.keycurriculum.com

First Edition Authors

Dan Fendel, Diane Resek, Lynne Alper, and Sherry Fraser

Contributors to the Second Edition

Sherry Fraser, Jean Klanica, Brian Lawler, Eric Robinson, Lew Romagnano, Rick Marks, Dan Brutlag, Alan Olds, Mike Bryant, Jeri P. Philbrick, Lori Green, Matt Bremer, Margaret DeArmond

Editor

Josephine Noah

Editorial Assistant

Emily Reed

Professional Reviewer

Rick Marks, Sonoma State University

Math Checker

Carrie Gongaware

Production Director

Christine Osborne

Production Editor

Andrew Jones

Executive Editor

Josephine Noah

Mathematics Product Manager

Elizabeth DeCarli

Publisher

Steven Rasmussen

Contents

Introduction

Activity Notes

Small World, Isn't It?

Intent

In this unit, students solve a problem involving population growth by fitting a function to a set of data. In preparation for this, they consider the nature of various mathematical descriptions of growth, including linear and exponential functions, slope, and derivatives. Students also learn about common and natural logarithms.

Mathematics

The main concepts and skills that students will encounter and practice during this unit are:

Rate of Change

- Evaluating average rate of change in terms of the coordinates of points on a graph
- Understanding the relationship between the rate of change of a function and the appearance of its graph
- Realizing that in many contexts, the rate of growth or decline with respect to time in a population is proportional to the population

Slope and Linear Functions

- Developing an algebraic definition of slope
- Proving, using similarity, that a line has a constant slope
- Understanding the significance of a negative slope for a graph and an applied context
- Seeing that the slope of a line is equal to the coefficient of x in the $y = a + bx$ representation of the line
- Using slope to develop equations for lines

Derivatives

- Developing the concept of the derivative of a function at a point
- Seeing that the derivative of a function at a point is the slope of the tangent line at that point
- Finding numerical estimates for the derivatives of functions at specific points
- Working with the derivative of a function as a function in itself
- Realizing that for functions of the form $y = b^x$, the derivative at each point of the graph is proportional to the y-value at that point

Exponential and Logarithmic Functions

- Using exponential functions to model real-life situations
- Strengthening understanding of logarithms
- Reviewing and applying the principles that $a^b \cdot a^c = a^{b+c}$ and $(a^b)^c = a^{bc}$
- Understanding and using the fact that $a^{\log_a b} = b$
- Discovering that any exponential function can be expressed using any positive number other than 1 as a base
- Learning the meaning of the terms *natural logarithm* and *common logarithm*
- Using an exponential function to fit a curve to numerical data

The Number e and Compound Interest

- Estimating the value of b for which the function $y = b^x$ has a derivative at each point on its graph equal to the y-value at that point
- Developing and using a formula for compound interest
- Seeing that expressions of the form $\left(1 + \dfrac{1}{n}\right)^n$ have a limiting value, called e, as n increases without bound
- Learning that the limiting value e is the same number as the special base for exponential functions

Students will work with other concepts in connection with the unit's Problems of the Week.

Progression

In *As the World Grows,* the unit opens with a table of world population data over the past several centuries and asks this question:

> *If population growth continues according to its current pattern, how long will it be until people are squashed up against one another?*

To answer this question, students begin with a variety of problems concerning rates of growth, focusing on the idea of an average rate of change, in *Average Growth*.

In *All In a Row,* the unit then focuses specifically on the case of constant change, represented by linear functions. Students develop the concept of slope, seeing that the rate of change can be represented by a ratio of coordinate differences. They then apply the concept of slope to develop equations for linear functions, based either on the slope and a point or on two points. They see that the slope of a straight line is equal to the coefficient of x in the algebraic function representation of the line. Students also use the concept of similarity to see why the slope of a straight line does not depend on the particular points used to compute the slope.

After this work with linear functions, in *Beyond Linearity* students examine nonlinear functions, looking for an analogous concept. Working with the notion of instantaneous velocity, they develop the concept of a derivative. This crucial concept is examined from several perspectives:

- *In terms of real-life situations:* Through examples such as the speed of a falling object, students connect the abstract concept of a derivative with more concrete models.
- *Numerically:* Students estimate derivatives numerically by computing the slope of the segment connecting "close" points on the graph.
- *Graphically:* Using a graphing calculator, students see that almost any graph they can create will look like a straight line if they zoom in close enough. This leads to discussion of the line tangent to a graph.

In *A Model for Population Growth,* students then examine two situations that fit a model of exponential growth—an inflation problem and a problem involving growth of an amoeba-like population. This leads to an examination of the derivatives of exponential functions, and students see that every exponential function has a derivative proportional to its y-value. This discovery fits with an intuitive analysis of population growth suggesting that the rate (with respect to time) at which a population is increasing should be proportional to the current population. Thus, exponential growth is a natural model to consider in solving the central unit problem.

In the next segment of the unit, *The Best Base,* students apply their understanding of bases and exponents to see that any positive number except 1 can be used as the base for any exponential function. Students estimate a value for the special base b for which the function $y = b^x$ has a derivative that is actually equal to its y-value. Then they study compound interest, recognizing that the number e that comes out of the compounding problem is the same number as the special base for the exponential function.

Ultimately, in *Back to the Data,* students return to the original population problem and try to fit an exponential function to the data they were given at the beginning of the unit.

As the World Grows

Average Growth

All In a Row

Beyond Linearity

A Model for Population Growth

The Best Base

Back to the Data

Pacing Guides

50-Minute Pacing Guide (34 days)

Day	Activity	Time Estimate
1	*As the World Grows*	0
	A Crowded Place	40
	Introduce: *POW 7: The More, the Merrier?*	10
	Homework: *How Many of Us Can Fit?*	0
2	Discussion: *How Many of Us Can Fit?*	10
	How Many More People?	40
	Homework: *Growing Up*	0
3	Discussion: *Growing Up*	10
	Discussion: *How Many More People?*	40
	Average Growth	0
	Homework: *Story Sketches*	0
4	Discussion: *Story Sketches*	10
	What a Mess!	40
	Homework: *Traveling Time*	0
5	Discussion: *Traveling Time*	10
	Comparative Growth	40
	Homework: *If Looks Don't Matter, What Does?*	0
6	Discussion: *If Looks Don't Matter, What Does?*	10
	All In a Row	0
	Formulating the Rate	40
	Homework: *Rates, Graphs, Slopes, and Equations*	0
7	Discussion: *Rates, Graphs, Slopes. and Equations*	10
	More About Tyler's Friends	40
	Homework: *Wake Up!*	0
8	Discussion: *Wake Up!*	40
	Discussion: *POW 7: The More the Merrier?*	10
	Homework: *California, Here I Come!*	0
9	Discussion: *California, Here I Come!*	30
	Introduce: *POW 8: Planning the Platforms*	20

	Homework: *Points, Slopes, and Equations*	0
10	Discussion: *Points, Slopes, and Equations*	10
	The Why of the Line	40
	Homework: *To the Rescue*	0
11	*Beyond Linearity*	0
	Discussion: *To the Rescue*	15
	The Instant of Impact	35
	Homework: *Doctor's Orders*	0
12	Discussion: *Doctor's Orders*	10
	Photo Finish	40
	Homework: *Speed and Slope*	0
13	Discussion: *Speed and Slope*	10
	ZOOOOOOOOM	40
	Homework: *The Growth of the Oil Slick*	0
14	Discussion: *The Growth of the Oil Slick*	15
	Discussion: *ZOOOOOOOOM*	25
	Homework: *Speeds, Rates, and Derivatives*	10
15	Discussion: *Speeds, Rates, and Derivatives*	15
	Zooming Free-For-All	35
	Homework: *On a Tangent*	0
16	Discussion: *On a Tangent*	10
	Presentations: *POW 8: Planning the Platforms*	25
	Introduce: *POW 9: Around King Arthur's Table*	15
	Homework: *What's It All About?*	0
17	Discussion: *What's It All About?*	15
	A Model for Population Growth	0
	How Much for Broken Eggs	35
	Homework: *Small but Plentiful*	0
18	Discussion: *How Much for Broken Eggs?!!?*	15
	Discussion: *Small but Plentiful*	30
	Homework: *The Return of Alice*	5
19	Discussion: *The Return of Alice*	10
	Slippery Slopes	40
	Homework: *The Forgotten Account*	0
20	Discussion: *The Forgotten Account*	10

	Slippery Slopes (continued)	40
	Homework: *How Does It Grow?*	0
21	Discussion: *Slippery Slopes*	25
	Discussion: *How Does It Grow?*	25
	Homework: *The Significance of a Sign*	0
22	Discussion: *The Significance of a Sign*	20
	Presentations: *POW 9: Around King Arthur's Table*	30
	Homework: *The Sound of a Logarithm*	0
23	Discussion: *The Sound of a Logarithm*	10
	The Power of Powers	40
	Homework: *The Power of Powers, Continued*	0
24	Discussion: *The Power of Powers, Continued*	20
	The Best Base	0
	A Basis for Disguise	30
	Homework: *Blue Book*	0
25	Discussion: *Blue Book*	20
	Discussion: *A Basis for Disguise*	20
	Homework: *California and Exponents*	10
26	Discussion: *California and Exponents*	10
	Find That Base!	40
	Homework: *Double Trouble*	0
27	Discussion: *Double Trouble*	15
	The Generous Banker	25
	Homework: *Comparing Derivatives*	10
28	Discussion: *Comparing Derivatives*	25
	Discussion: *The Generous Banker*	25
	Homework: *The Limit of Their Generosity*	0
29	Discussion: *The Limit of Their Generosity*	50
	Homework: *California Population with e's*	0
30	Discussion: *California Population with e's*	10
	Back to the Data	0
	Tweaking the Function	40
	Homework: *Beginning Portfolios–Part I*	0
31	Discussion: *Beginning Portfolios–Part I*	10
	Return to "A Crowded Place"	40

		Homework: *Beginning Portfolios–Part II*	0
32		Discussion: *Beginning Portfolios–Part II*	10
		Presentations: *Return to "A Crowded Place"*	40
		Homework: *"Small World, Isn't It?" Portfolio*	0
33		*In-Class Assessment*	30
		Homework: *Take-Home Assessment*	20
34		Exam Discussion	35
		Unit Reflection	15

90-Minute Pacing Guide (22 days)

Day	Activity	Time Estimate
1	*As the World Grows*	0
	A Crowded Place	40
	How Many More People?	40
	Introduce: *POW 7: The More, the Merrier?*	10
	Homework: *How Many of Us Can Fit?*	0
2	Discussion: *How Many of Us Can Fit?*	10
	Discussion: *How Many More People?*	40
	Growing Up	40
	Average Growth	0
	Homework: *Story Sketches*	0
3	Discussion: *Story Sketches*	10
	What a Mess!	40
	Comparative Growth	40
	Homework: *Traveling Time*	0
4	Discussion: *Traveling Time*	10
	If Looks Don't Matter, What Does?	35
	All In a Row	0
	Formulating the Rate	45
	Homework: *Rates, Graphs, Slopes, and Equations*	0
5	Discussion: *Rates, Graphs, Slopes. and Equations*	15
	More About Tyler's Friends	40
	Wake Up!	35
	Homework: *California, Here I Come!*	0
6	Discussion: *California, Here I Come!*	20
	Discussion: *Wake Up!*	30
	The Why of the Line	40
	Homework: *Points, Slopes, and Equations*	0
7	Discussion: *Points, Slopes, and Equations*	10
	Beyond Linearity	0

	To the Rescue	45
	The Instant of Impact	35
	Homework: *Doctor's Orders*	0
8	Discussion: *Doctor's Orders*	10
	Discussion: *POW 7: The More the Merrier?*	15
	Introduce: *POW 8: Planning the Platforms*	20
	Photo Finish	45
	Homework: *Speed and Slope*	0
9	Discussion: *Speed and Slope*	10
	ZOOOOOOOOM	65
	Homework: *The Growth of the Oil Slick*	15
10	Discussion: *The Growth of the Oil Slick*	10
	Speeds, Rates, and Derivatives	50
	Zooming Free-For-All	30
	Homework: *On a Tangent*	0
11	Discussion: *On a Tangent*	10
	A Model for Population Growth	0
	How Much for Broken Eggs?!!?	50
	Small but Plentiful	30
	Homework: *What's It All About?*	0
12	Discussion: *What's It All About?*	15
	Discussion: *Small but Plentiful*	30
	The Return of Alice	45
	Homework: *The Forgotten Account*	0
13	Discussion: *The Forgotten Account*	10
	Slippery Slopes	80
	Homework: *How Does It Grow?*	0
14	Discussion: *Slippery Slopes*	25
	Discussion: *How Does It Grow?*	25
	The Significance of a Sign	40
	Homework: *The Sound of a Logarithm*	0
15	Discussion: *The Sound of a Logarithm*	10
	Presentations: *POW 8: Planning the Platforms*	25
	Introduce: *POW 9: Around King Arthur's Table*	15
	The Power of Powers	40

	Homework: *The Power of Powers, Continued*	0
16	Discussion: *The Power of Powers, Continued*	20
	The Best Base	0
	A Basis for Disguise	50
	Homework: *California and Exponents*	20
17	Discussion: *California and Exponents*	10
	Blue Book	40
	Find That Base!	40
	Homework: *Double Trouble*	0
18	Discussion: *Double Trouble*	15
	The Generous Banker	45
	The Limit of Their Generosity	25
	Homework: *Comparing Derivatives*	5
19	Discussion: *Comparing Derivatives*	15
	Discussion: *The Limit of Their Generosity*	50
	California Population with e's	25
	Homework: *Beginning Portfolios–Part I*	0
20	Discussion: *Beginning Portfolios–Part I*	5
	Discussion: *California Population with e's*	10
	Back to the Data	0
	Tweaking the Function	40
	Return to "A Crowded Place"	35
	Homework: *Take-Home Assessment*	0
21	Presentations: *Return to "A Crowded Place"*	45
	In-Class Assessment	30
	Homework: *Beginning Portfolios–Part II*	15
22	Presentations: *POW 9: Around King Arthur's Table*	25
	Exam Discussion	45
	Discussion: *Beginning Portfolios–Part II*	5
	Unit Reflection	15
	Homework: *"Small World, Isn't It?" Portfolio*	0

Materials and Supplies

All IMP classrooms should have a set of standard supplies, described in the section "Materials and Supplies for the IMP Classroom" in *A Guide to IMP.* You'll also find a comprehensive list of materials needed for all Year 3 units in the section "Materials and Supplies for Year 3" in the *Year 3 Teacher's Guide* general resources.

No additional supplies are needed for this unit. However, general and activity-specific blackline masters, for transparencies or for student worksheets, are available in the "Blackline Masters" section in *Small World, Isn't It?* Unit Resources.

More About Supplies

Graph paper is a standard supply for IMP classrooms. Blackline masters of 1-Centimeter Graph Paper, ¼-Inch Graph Paper, and 1-Inch Graph Paper are provided, for you to make copies and transparencies.

Assessing Progress

Small World, Isn't It? concludes with two formal unit assessments. In addition, there are many opportunities for more informal, ongoing assessments throughout the unit. For more information about assessment and grading, including general information about the end-of-unit assessments and how to use them, consult *A Guide to IMP*.

End-of-Unit Assessments

This unit concludes with in-class and take-home assessments. The in-class assessment is intentionally short so that time pressures will not affect student performance. Students may use graphing calculators and their notes from previous work when they take the assessments. You can download unit assessments from the *Small World, Isn't It* Unit Resources.

Ongoing Assessment

One of the primary tasks of the classroom teacher is to assess student learning. Although the assigning of course grades may be part of this process, assessment more broadly includes the daily work of determining how well students understand key ideas and what level of achievement they have attained on key skills, in order to provide the best possible ongoing instructional program for them.

Students' written and oral work provides many opportunities for teachers to gather this information. Here are some recommendations of written assignments and oral presentations to monitor especially carefully that will give you insight into student progress:

- *How Many More People?*
- *Points, Slopes, and Equations*
- *Photo Finish*
- *What's It All About?*
- *Slippery Slopes*
- *Return to "A Crowded Place"*

Discussion of Unit Assessments

Ask for volunteers to explain their work on each of the problems. Encourage questions and alternate explanations from other students.

In-Class Assessment

Question 1 covers the key ideas about equations of straight lines. Students may have a variety of approaches for developing the formula in Question 1a. Questions 1b and 1c will demonstrate their understanding of the relationship between the formula and the situation.

Similarly, Question 2 covers the definition and significance of the derivative.

Take-Home Assessment

Question 1 deals with derivatives from a graphical perspective. Question 1a is straightforward graph reading, but Question 1b will require students to use something like the idea of a tangent line. Question 1c reviews basic principles about the sign of the derivative.

Students' responses to Question 2 should give you a good idea of how well they understood the difference between absolute growth rate and proportional growth rate and the relationship between population growth and the proportionality property of exponential functions.

Supplemental Activities

Small World, Isn't It? contains a variety of activities at the end of the student pages that you can use to supplement the regular unit material. These activities fall roughly into two categories.

Reinforcements increase students' understanding of and comfort with concepts, techniques, and methods that are discussed in class and are central to the unit.

Extensions allow students to explore ideas beyond those presented in the unit, including generalizations and abstractions of ideas.

The supplemental activities are presented in the *Teacher's Guide* and the student book in the approximate sequence in which you might use them. Listed here are specific recommendations about how each activity might work within the unit. You may wish to use some of these activities, especially the later ones, after the unit is completed.

Solving for Slope **(reinforcement or extension)** This activity involves finding the slope algebraically when a linear equation does not give y in terms of x. It can be assigned after students recognize that the slope of a linear equation in the form $y = ax + b$ is equal to the coefficient a. (They will probably see this in the discussion of *Rates, Graphs, Slopes, and Formulas*.)

Slope and Slant **(extension)** This activity introduces the concept of *angle of inclination* and asks students to investigate the relationship between this angle and the slope of a line. The activity fits into the unit following the discussion of *Rates, Graphs, Slopes, and Formulas*.

Predicting Parallels **(extension)** This activity continues the theme of *Solving for Slope* and can be assigned at the same time as that supplemental activity.

The Slope's the Thing **(extension)** In this activity, students investigate a standard procedure for developing the equation of a line from the coordinates of two of its points. This work makes a good follow-up to the discussion of *The Why of the Line*.

Speedy's Speed by Algebra **(extension)** This activity introduces students to the algebraic approach to derivatives that is commonly used in calculus classes (although the activity does not explicitly refer to derivatives). You can suggest that students work on this after the discussion of *What's It All About?*

Potential Disaster (reinforcement) This activity is similar to the oil slick problems (*What A Mess!* and *The Growth of the Oil Slick*) in that it describes the modeling of a growth situation using an equation. You can assign it anytime after *The Growth of the Oil Slick*.

Proving the Tangent (extension) In Question 1 of *On a Tangent,* students find the derivative of the function $f(x) = 0.5x^2$ at the point (2, 2) and connect the value of the derivative to the slope of the tangent line. This activity follows up on that activity by having students use the derivative to get the equation of the tangent line and show that this line meets the graph in only one point.

Summing the Sequences—Part I (reinforcement) This activity explores the sums of arithmetic sequences. If students did not get a closed-form expression for the total length of material needed in *POW 8: Planning the Platforms,* this activity provides an opportunity for them to do so.

Summing the Sequences—Part II (extension) The concept of a geometric sequence was introduced in the supplemental activity *More About Rallods* in the Year 2 unit *All About Alice.* This activity gives students another opportunity to work with this concept and is a natural follow-up to *Summing the Sequences—Part I.*

Looking at Logarithms (extension) In this activity, students are asked to find logarithm analogs of some general principles about exponents. Students can work on this after the review of logarithms in *The Return of Alice.*

Finding a Function (extension) This activity extends the idea introduced in *The Significance of a Sign* about finding functions whose graphs fit specific conditions. You can use it as a follow-up to that activity.

Deriving Derivatives (extension) This activity could be assigned anytime after the definition of derivative in the discussion introducing *Speeds, Rates, and Derivatives.* But students may have more success if you wait until after *Slippery Slopes,* in which they get experience making tables of derivative values, and after the discussion of treating the derivative as a function that follows *The Significance of a Sign.*

The Reality of Compounding (reinforcement) This activity is a follow-up to the discussion of *The Generous Banker.*

Transcendental Numbers (extension) This activity can be assigned anytime after *California Population with e's.*

Dr. Doubleday's Base (reinforcement) This activity gives students a further opportunity to work concretely with the proportionality property that they discovered in *Slippery Slopes*. It should be used after the definition of *e*, so could be assigned about the same time as *California Population with e's*.

Investigating Constants (extension) Like *Dr. Doubleday's Base,* this activity is an exploration of the proportionality property, but it is posed in a more open-ended way and should be considered more challenging than *Dr. Doubleday's Base*. It can be assigned after *California Population with e's*.

As The World Grows

Intent

This sequence of activities introduces the central unit problem.

Mathematics

In the activities of *As The World Grows*, students explore the relationship between graphs and rates of change. They calculate and compare average rates of change and recognize the distinction between amount of increase and percent of increase.

Progression

A Crowded Place introduces the central unit problem, which is to use world population data from the past several centuries to predict when the people of the world will be "all squashed up against one another." *How Many of Us Can Fit?* refines this question by allotting each person a single square foot of dry land.

In *How Many More People?* and *Growing Up*, students look at rate of change while working with graphs.

A Crowded Place

POW 7: The More, the Merrier?

How Many of Us Can Fit?

How Many More People?

Growing Up

A Crowded Place

Intent
This activity introduces the central unit problem.

Mathematics
A Crowded Place presents students with world population data for the past several centuries. The activity asks students to predict, if this pattern of data continued, how long it would take until we were "all squashed up against one another." The problem also provides the earth's surface area and the percentage of that area that is land.

As students brainstorm criteria that might affect population growth, they examine data with the goal of predicting future behavior.

Progression
Students will return to the unit question at various points throughout the unit. The focus in this activity is on getting students engaged with the issue, and the bulk of class time should be spent in discussion rather than on computations.

Approximate Time
20 minutes for activity

20 minutes for discussion

Classroom Organization
Small groups, followed by whole-class discussion

Doing the Activity
Have groups read the title activity *A Crowded Place* and give them some time to explore the data and the question. Bring the class together for discussion after about 15 minutes of work on the activity.

Although some groups may come up with numerical answers, the primary goals are to get students interested in the question and to have them begin thinking about the issues involved in answering it. For instance, they should realize that the phrase "squashed up against one another" is ambiguous and will need to be defined more carefully in order to answer the question. They may wonder about how to deal with people in multi-storied buildings (people above one another) and people on ships at sea. Encourage these musings—they add to the intrigue of the problem. (In *How*

Many of Us Can Fit?, the phrase "squashed up against one another" is interpreted to mean that there is one square foot of land per person.)

Students may be tempted to turn immediately to a graph or their calculators to plot the points and look for a function. That's a good first approach, but tell them that for the next few weeks they will be looking at some new ideas that will give them further insight into the way population grows.

Groups will probably realize that the question itself is rather facetious. Long before people could get "squashed up against one another," the impact of overcrowding would affect the population growth rate. As with other units that students have studied, the initial problem is concrete but ambiguous, and students need to make some simplifying assumptions in order to make progress.

Students will work further with this data set in *How Many More People?* and occasionally over the first couple of weeks of the unit. In *Return to A Crowded Place*, they will return to the data to try to answer the central question.

Discussing and Debriefing the Activity

Bring the class together and let groups briefly share any numerical results they came up with. But focus the discussion on the issues that need to get resolved, rather than on finding a specific answer to the question.

To put the problem in a broader context, ask, **What factors would affect the rate at which a population grows?** Students should come up with some items from this list as possible factors:

- medical advances
- limitations on resources such as food and energy
- birth control policies and practices
- war
- environmental concerns
- space travel
- cultural values about family size

For each item mentioned, ask, **Would this item increase or decrease the rate of population growth?** Students should be realizing that these factors are unpredictable, and so their final answer to the question raised in *A Crowded Place* will be very speculative.

Ask, **Why are people concerned about excessive population growth?** The first POW of the unit, *POW 7: The More, the Merrier?,* asks students to write reports on some aspect of population growth, and you may want to save suggestions made here as sources of POW topics.

Key Questions

What factors would affect the rate at which a population grows?

Would this item increase or decrease the rate of population growth?

Why are people concerned about excessive population growth?

POW 7: The More, the Merrier?

Intent

This activity is intended to increase students' interest and engagement in the central unit problem.

Mathematics

The More the Merrier? gives students the opportunity to explore some of the social implications of population growth that will naturally arise as they consider the unit problem, so it will cause them to become more engaged with that question. It will also help them to understand the need for simplifying assumptions, since population growth rates are shaped by social issues, and cannot be realistically predicted by curve fitting alone.

Progression

The activity asks students to research and report on some aspect of the concern over rapid population growth. The teacher introduces the activity with some discussion of expectations for the research paper. Students should be given sufficient time to locate research materials. Unlike most POW's, no student presentations are needed in this case.

Approximate Time

10 minutes for introduction

1 to 3 hours for activity (at home)

10 to 15 minutes for discussion

Classroom Organization

Individuals, followed by whole-class discussion

Doing the Activity

As a class, go over rules that apply to research papers. For instance, make sure students realize that they are to use their own words unless they use quotation marks and cite their source. You may want to suggest or require that students use one or more sources other than the Internet, such as newspaper articles or encyclopedias.

Tell students that if they use the Internet as a source, they must list specific Web sites referenced. You may want to caution them that there is a great deal of misinformation on the Internet, so they should consider the likely reliability of what they find. For example, governmental Web sites tend to be more reliable than personal Web sites, or sites with a particular agenda.

Expectations and Priorities

Discuss with your class your expectations and priorities for this POW. For instance, you may want to emphasize original thinking, or you may want students to focus on clear organization, including a good summary. You also might specify a particular length that you want papers to be.

Discussing and Debriefing the Activity

Let students share what they learned with their fellow group members. As time allows, you may want to have some discussion of ideas as a whole class.

How Many of Us Can Fit?

Intent

In this activity, students work with a specific definition of "squashed up against one another" to refine the central unit problem.

Mathematics

The activity requires students to find the amount of land on the earth by using the percentage of the surface area given in *A Crowded Place*, and then to convert from square miles to square feet.

Progression

In this activity, students refine the unit problem by finding the number of square feet of land, and defining "squashed up against one another" to mean that each person has one square foot.

Approximate Time

15 minutes for activity (at home or in class)

10 minutes for discussion

Classroom Organization

Individuals, followed by whole-class discussion

Doing the Activity

Students work on this activity independently.

Discussing and Debriefing the Activity

Give groups a few minutes to compare results, and then let students from one or two groups report on each question.

Before going over the answers, you might dramatize the assumption in Question 4 by marking off a 1-foot–by–1-foot square and having a student stand within that space.

The problems are all direct computations. The answers are:

- Question 1: 29.2% of 196,930,000 square miles is approximately 57,500,000 square miles.

- Question 2: 57,500,000 square miles times 5280^2 square feet per square mile gives about 1.6×10^{15} (1.6 quadrillion) square feet.

- Question 3: 1.6×10^{15} square feet shared equally by 6.48 billion people yields about 247,000 square feet per person.

- Question 4: The number of people would be the same as the number of square feet of land area, so there would be about 1.6×10^{15} people.

You may want to let students speculate now on how long it would take before the population got to this size.

The Central Question

Use the result from Question 4 to redefine the central problem of the unit, and then post the new question: Based on the data in *A Crowded Place* how long will it take until the population reaches 1.6×10^{15} people?

Key Question

Based on the data in *A Crowded Place* how long will it take until the population reaches 1.6×10^{15} people?

How Many More People?

Intent

In this activity, students find an average rate of change and interpret change shown in a graph.

Mathematics

This activity explores the data from *A Crowded Place*, focusing on rates of population growth and how changes in those rates are reflected in a graph. The concept of **average rate of change** is introduced, in preparation for development of the concept of slope. The difference between discrete and continuous graphs is also noted.

Progression

Students graph the population data from *A Crowded Place* and then find and compare average population growth for several intervals. The discussion focuses on how the computations are reflected in the graph.

Approximate Time

40 minutes for activity

40 minutes for discussion

Classroom Organization

Small groups, followed by whole-class discussion

Materials

Transparency of *How Many More People?* blackline master

Doing the Activity

Have students work on this activity in groups. The initial task of drawing the graph will require them to choose an appropriate scale. You may want to point out to individual groups that they need not begin the horizontal (time) axis at the year 0. For instance, they might choose to begin at 1600 or 1650.

For use in the discussion of Part I, you might make a transparency of the graph provided in the *How Many More People?* blackline master, or have students make overhead transparencies of their own graphs. Even if you use the blackline master, you will still want to pass out overhead transparencies and pens to some groups so they can present their answers to the other questions of the activity.

Discussing and Debriefing the Activity

Part I: The Graph

The main issue here is the choice of an appropriate scale. You may want to display a transparency of the *How Many More People?* blackline master and have groups discuss how the graph's appearance would change due to differences in scale.

Ask, **Did you connect the points on your graph?** Remind students that a graph of isolated points is called *discrete*, and that connecting the points, whether by line segments or by smooth curves, yields a *continuous* graph.

Ask, **What assumption would be implied by connecting the points by line segments?** Bring out that doing so would imply an assumption that the population grew at a constant rate during each period. (This idea is discussed further in Part II.)

Part II: Average Increases

Have students use the transparency of the *How Many More People?* blackline master in their presentations for Part II. This will provide a standard for comparison as well as an accurately drawn graph from which they can work.

Have a student present his or her work for both Questions 2a and 2b, showing (for part b) how the numerical difference from part a appears on the graph. Geometrically, there are several ways for students to show this increase; elicit alternative approaches from the class.

Some students may indicate how the change shows up along the vertical (population) axis. Others may work with the points themselves, perhaps focusing on the vertical change from the point (1650, 470,000,000) to (1900, 1,550,000,000). Both are illustrated in this diagram:

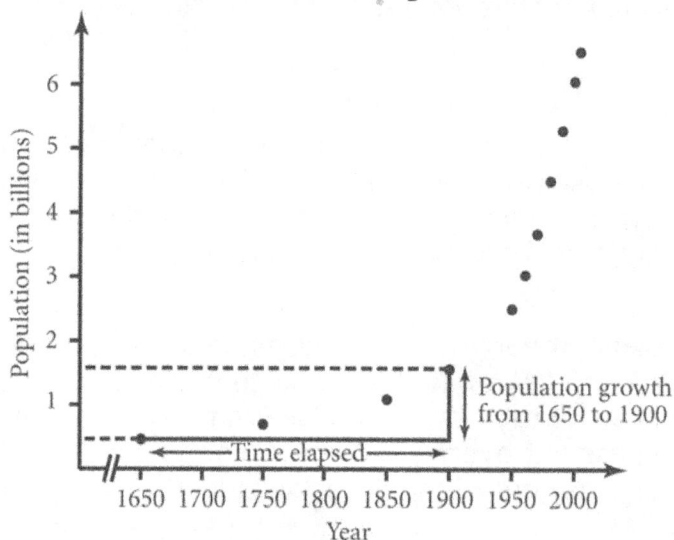

Then move on to parts c and d of Question 2. Ask the presenter of Question 2c to articulate the details of the computations so the class sees that the answer is the ratio of two differences:

$$\frac{\text{population in 1900} - \text{population in 1650}}{1900 - 1650}$$

This ratio is equal to about 4.3 million people per year.

Point out that this ratio can be expressed generically as

$$\frac{\text{change in population}}{\text{time elapsed}}$$

Describe the number calculated in this way as representing the **average rate of change** of the population over the given time interval. This idea will play an important role in the development of the concept of slope.

Ask, **What does "average" mean in this context?** Bring out that although the average rate of change for this time period is about 4.3 million people per year, there need not have been any one year that experienced exactly this amount of increase.

The key idea in Question 2d is that if the amount of population increase during the period 1650 to 1900 were the same every year, then that portion of the graph would be a line segment. Students can think of such a straight-line graph as taking the years of rapid growth and the years of slow growth and spreading the growth out evenly.

Don't press for a detailed explanation at this time for why a constant growth rate is associated with a straight-line graph. What is important now is establishing the intuitive connection. Students will examine the reasons more carefully later in the unit (see *The Why of the Line*).

Because Question 3 covers the same ideas as Question 2, you can use the presentations of Question 3 to solidify the ideas discussed in Question 2.

Part III: Making Comparisons

Finally, have one or two students give their ideas for Question 4. As part of this discussion, be sure that students draw the line segments connecting the data points for the ends of each interval, as shown here:

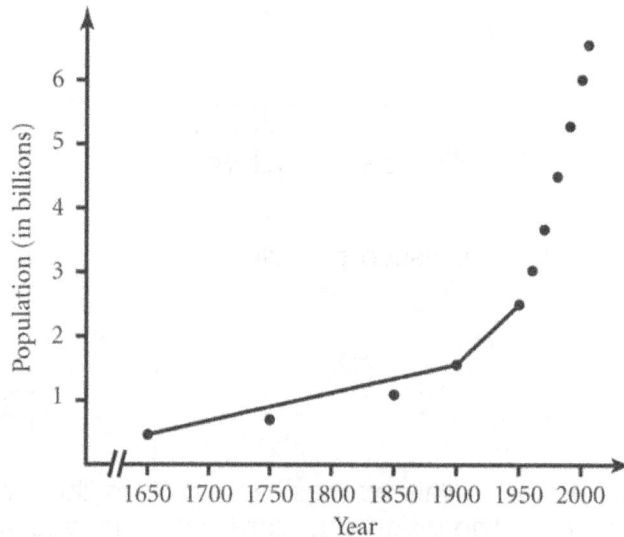

Students should begin to associate the steepness of a line with the rate of change it represents. For instance, most students should be able to articulate that they can see that there was a higher average population growth per year during the period from 1900 to 1950 than from 1650 to 1900, because the graph is steeper during the later time period.

Key Questions

Did you connect the points on your graph?

What assumption would be implied by connecting the points by line segments?

What does "average" mean in this context?

Growing Up

Intent

In this activity, students consider how rates of growth are reflected in a graph.

Mathematics

This activity provides further practice in reading and interpreting graphs. It also begins to develop the critical idea that growth or change can be measured in two different ways: as an absolute amount and as a proportional or percentage change.

Progression

Students answer questions based upon a graph of the average height of boys from birth to age 6. The discussion covers the distinction between amount of increase and percentage of increase, which will play a significant role in selecting a family of functions to model population growth for the unit problem.

Approximate Time

30 minutes for activity (at home or in class)

10 minutes for discussion

Classroom Organization

Individuals, followed by whole-class discussion

Materials

Transparency of *Growing Up* blackline master

Doing the Activity

You may need to remind students of the convention used on this graph of having a jagged line for part of the vertical axis to indicate that a portion of the axis is omitted.

Discussing and Debriefing the Activity

The questions in this activity are fairly straightforward, though there may be some variation in answers based on the degree of precision with which students were able to read the graph. The graph in this activity is provided as a blackline master.

Discuss this activity as a whole class, having students share their responses. In Question 2, students need only say that the near-straightness of the graph over this interval indicates a constant growth rate.

In Question 4, bring out that while the answers to Questions 3a and 4a are essentially equal, the answers for Questions 3b and 4b are not, because the same *amount* of height increase is a smaller *percentage* increase at age 5 than it was at age 3. Encourage students to begin to articulate the difference between these two ways of describing growth. This idea will be developed further in subsequent activities.

Average Growth

Intent

In this section, students look at average growth rates in preparation for studying slope.

Mathematics

These activities focus on the relationship between the steepness of a graph and rate of change. They prepare students to deal with several topics that are central to this unit: slope, derivatives (instantaneous rate), and algebraic representation of rates of change.

Progression

Story Sketches explores the connection between rate of change and the steepness of a graph, while *Comparative Growth* accents the importance of considering scale as well as steepness if comparing rates of growth using two graphs. In *What a Mess!* and *Traveling Time*, students work with rates without graphs. *If Looks Don't Matter, What Does?* sets the stage for a formal development of slope.

Story Sketches

What a Mess!

Traveling Time

Comparative Growth

If Looks Don't Matter, What Does?

Story Sketches

Intent

The purpose of this assignment is to give students more experience with the relationship between rate of change and steepness of graph for both straight lines and curves.

Mathematics

In this activity, students estimate a rate of change from a graph, relating the rate of change to the steepness of the graph. They see the usefulness of simplifying a mathematical model by using a linear graph in place of a step function and see that functions whose graphs are parallel lines have equal rates of change.

Progression

Students work on the activity individually and then discuss their results as a class.

Students will further explore some of these graphs in *More About Tyler's Friends,* so you may want to suggest that they save their work.

Approximate Time

30 minutes for activity (at home or in class)

10 minutes for discussion

Classroom Organization

Individuals, followed by whole-class discussion

Materials

Transparency of *Story Sketches* blackline master

Doing the Activity

Students work on this activity independently.

Discussing and Debriefing the Activity

The graph in Question 1 is provided in the *Story Sketches* blackline master.

Let students present different components of Questions 1 and 2. In both situations, emphasize that they are looking at rates of growth: of the distance from Westport; and of the amounts of money in the piggy banks. Point out as needed that "rate of growth" generally measures how something changes over time, so it's appropriate that in both situations, the horizontal axis of the graph represents time.

Question 1

In Question 1, students need to make numerical estimates based on the graph. This may not be easy, because the graph is small, but don't be too concerned about the accuracy of the estimates (perhaps about 4 miles per day for part a, and about 15-20 miles per day for part b). Emphasize the similarity between the arithmetic involved in these questions and students' work on *How Many More People?*

Even if students have difficulty getting accurate estimates for Question 1b, they should realize that they should focus on the steepest part of the graph.

Question 2

On Question 2, students have several options in drawing the graphs. One likely choice is to use discrete graphs. For instance, the graph for Tyler might look like this:

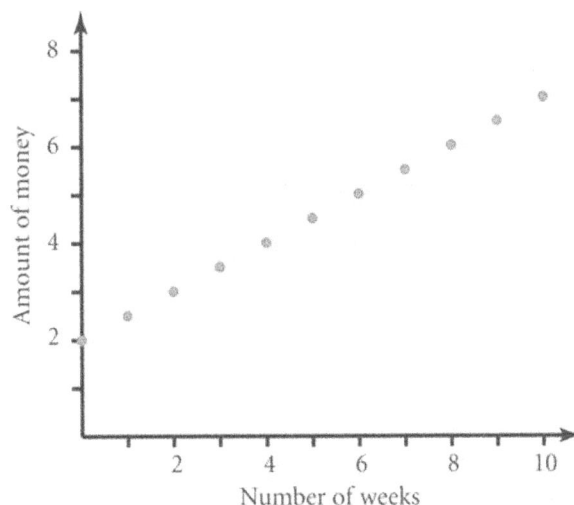

Another likely choice is to connect the individual points with a straight line. Thus, one might represent Tyler's situation with this graph:

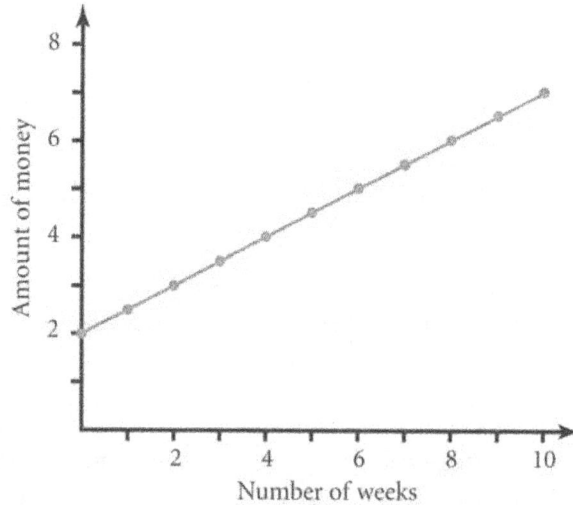

Note that this is a simplification, though a reasonable one. Get the class to articulate that this graph would indicate that the allowance is given out gradually throughout the week.

A Step Function

Each child's accumulated savings is more accurately described by a step function, since the amount in the piggy bank only changes at the end of each week. For example, it's more accurate to draw Tyler's graph like this:

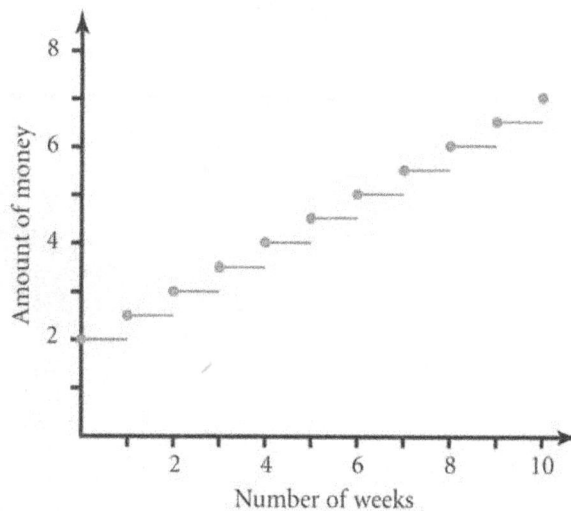

Bring out that such a graph is a better description of what's happening than a discrete graph, and introduce the term **step function** for a graph like this.

Though you should definitely discuss the step function approach, tell students that for the discussion of Question 3, they should imagine that the money is being added gradually, and thus connect the points with a straight line. Note also that it's easier to compare the children to one another using straight-line graphs than using step functions. If necessary, have the class redo their graphs in this manner, but reiterate that this is a simplification.

Question 3

In discussing Question 3 (based on the straight-line versions of the graphs), be sure students see there are two key ways in which the graphs differ from one another:

- Some of the graphs have different "steepnesses," which reflect the different allowance rates.
- Some of the graphs have different "starting points," which reflect the different amounts the children had initially.

Ask, **How do the graphs show that Robin and Max had the same allowance?** Students should see that their graphs are parallel lines. Also ask, **How do the graphs show that Tyler and Robin started with the same amount?** Students should note that the graphs for these two children intersect the vertical axis at the same point. Ask, **What do we call the place where a graph crosses the vertical axis?** If needed, remind the class of the term *y-intercept*.

Question 4

You might have volunteers give equations for each child's situation. For instance, students might describe the amount in Tyler's piggy bank with the equation $m = 2 + 0.50t$ (in which m represents the amount of money and t the number of weeks elapsed). Bring out that this is a linear equation, which gives another reason for representing the situation by a straight line.

Key Questions

How do the graphs show that Robin and Max had the same allowance?

How do the graphs show that Tyler and Robin started with the same amount?

What do we call the place where a graph crosses the vertical axis?

What a Mess!

Intent

In this activity, students use linear expressions to study the problem of cleaning up an oil spill.

Mathematics

The activity describes a circular oil spill whose radius is increasing at a constant rate. Students make a table, draw a graph, and write a rule to represent this situation. They then compare another related constant rate.

Progression

This activity is one of several in which students explore the relationship between rate of change and the equation of a straight line. They will use these examples as the basis for developing the concept of slope as a constant rate of change.

Approximate Time

30 minutes for activity

10 minutes for discussion

Classroom Organization

Small groups, followed by whole-class discussion

Doing the Activity

You may want to suggest to groups that they use $t = 0$ to represent the time when Lindsay first sees the oil slick.

Discussing and Debriefing the Activity

Questions 1, 2, and 3

Let different students present the In-Out table, the graph, and the rule. They should find a rule equivalent to $r = 70 + 6t$, relating the radius to the time elapsed. Be sure that students articulate in the discussion that the coefficient 6 in the expression $6t$ represents the number of meters per hour at which r is growing. (Students will revisit the rule $r = 70 + 6t$ in *The Growth of the Oil Slick*.)

If the presenter of the graph does not label the axes clearly, ask, **How should the axes be labeled?** Both axes should include labels and scales. (Because the values of r start at 70, students may wish to skip a segment of the vertical axis, as shown here.)

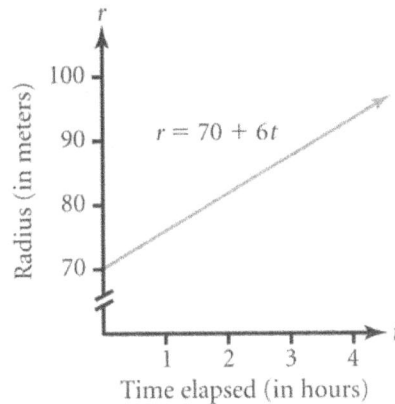

Also, be sure that students realize that the graph of the situation is not just a set of discrete points, but that all the in-between values work as well. That is, the *In* value—the number of hours—can be a fraction as well as a whole number.

Question 4

Elicit a variety of approaches for Question 4. If students are having trouble, one approach is to set up an In-Out table giving the radius for both the oil-slick circle and the clean-water circle for various values of *t*. Such a table might look like this:

Number of hours since the cleanup began	Radius of the oil slick, in meters	Radius of the clean-water circle, in meters
0	112	0
1	118	10
2	124	20
8	160	80

This table may serve as a lead-in to a traditional algebraic approach, in which students develop equations for the radius of each circle *t* hours after the clean-up began. The oil slick has a radius of 112 meters when the pumping of detergent begins, and increases at 6 meters per hour. Thus, after *t* hours, the radius of the oil slick is $112 + 6t$ meters. The clean-water circle has radius 0 when the pumping begins, and increases at 10 meters per hour; so after *t* hours, this circle has a radius of $10t$ meters.

The table might help students see that they want these two radii to be equal. In algebraic terms, this occurs when

$$112 + 6t = 10t$$

This equation simplifies to $112 = 4t$, so the oil slick will be completely eliminated after 28 hours.

Another approach to solving Question 4 is to use the fact that the radius of the clean-water circle is "gaining" on the radius of the oil slick at a rate of 4 meters per hour. Since the radius of the clean-water circle is initially 112 meters less than that of the oil slick, it will take 112 ÷ 4 hours for the oil slick to be completely eliminated.

Key Question

How should the axes be labeled?

Traveling Time

Intent

In this activity, students continue their work with rates.

Mathematics

This activity gives students further practice working with rates. It also provides an opportunity to explore one of the common mistakes in working with average rates—that of averaging the speed for periods in which an equal *distance* was traveled, rather than for periods of equal time. Finally, it provides an initial exposure to the distinction between average rate and instantaneous rate.

Progression

Students work through various rate problems, ranging from straightforward to more complex.

Approximate Time

30 minutes for activity (at home or in class)

10 minutes for discussion

Classroom Organization

Individuals, followed by whole-class discussion

Materials

Transparency of *Traveling Time* blackline master

Doing the Activity

The first question in *Traveling Time* is a very straightforward rate problem. The second question requires more thought and several calculations, presenting a situation in which a given rate of travel is averaged for a portion of a trip and asking what speed must be averaged for the remainder of the trip to achieve a particular arrival time.

Question 3 will catch many students in a trap, as it asks them to make up their own scenario in which a train averaged 50 mph for a trip, but traveled at least two different speeds along the way. Simply choosing two speeds that average to 50 mph, such as 40 and 60 mph, only works under certain conditions.

The final question asks students to compare rates of travel for different time periods and raises the issue of instantaneous rates.

Discussing and Debriefing the Activity

Assign groups to prepare presentations on each problem. (Question 3 lends itself well to multiple presentations.) You might provide a transparency of the graph from Question 4, which is included in the *Traveling Time* blackline master, so that the groups presenting that question need not draw the graph on a transparency.

Question 1 is fairly straightforward and you may only want to have one group present it. Students should see that the train has 10 hours to cover the 550 miles, so it must average 55 miles per hour.

For Question 2, students will need to figure out how far the train has traveled in the first hour-and-a-half (60 miles), calculate the remaining distance (490 miles) and remaining time ($8\frac{1}{2}$ hours), and then find the speed required (approximately 57.6 miles per hour).

Question 3

For Question, be sure to have the class check the scenarios presented for creating an overall average speed of 50 miles per hour, since it is easy to fall into a trap on this problem.

For instance, it's common to suggest having the train travel the first 275 miles at 40 miles per hour and the final 275 miles at 60 miles per hour. However, under this scenario, the train only averages 48 miles per hour, and the trip takes about a half-hour longer than it would if the train went 50 miles per hour for the whole trip. The trap is that in order for speeds to "average," the train needs to travel the same amount of *time* at each speed, not travel the same *distance* at each speed.

Question 4

Question 4 serves two purposes: it gives students another opportunity to work with the steepness of a graph and how it relates to the average rate of change; and , it is an important initial exposure to the difference between average and instantaneous speed.

Comparative Growth

Intent

In this activity, students observe that the steepness of a graph is not always an adequate measure of rate of change.

Mathematics

Students calculate the average growth rate from two graphs of population that have different scales. They realize that both steepness and scale must be taken into consideration when comparing rates.

Progression

Students calculate average growth using two graphs, and observe that the steeper graph does not necessarily represent the greater rate of change, if the graphs are not scaled alike.

Approximate Time

30 minutes for activity

10 minutes for discussion

Classroom Organization

Small groups, followed by whole-class discussion

Materials

Optional: Transparency of *Comparative Growth* blackline master

Doing the Activity

Students work on this activity independently.

Discussing and Debriefing the Activity

Have students share their results with their groups. When most groups have reached consensus, ask a student from each group to report on the answers to Questions 1 and 2. Be sure they articulate the arithmetic used to answer the question. Discuss any disagreement on these numerical results (beyond slight variations due to different estimates from the graph).

Continue by having a couple of groups report on Question 3. The important observation is that, although the first graph generally looks steeper, the scales of the graphs are different, and so the visual appearance is not a reliable basis for comparison of steepness. The *Comparative Growth* blackline master provides a single graph using a common scale that combines the information from both graphs. You may want to use a transparency of this blackline master in the discussion of Question 3.

If Looks Don't Matter, What Does?

Intent

The discussion of this activity leads to the formal definition of slope.

Mathematics

This activity focuses on the process for computing the average rate of change using coordinates.

Progression

In the previous activity, *Comparative Growth,* students learned that steepness cannot be accurately assessed visually. In this activity, they compute average rate of change using coordinates. In the following activity, *Formulating the Rate*, students develop the formula for slope.

Approximate Time

25 minutes for activity (at home or in class)

10 minutes for discussion

Classroom Organization

Individuals, followed by whole-class discussion

Doing the Activity

Students work on this activity independently.

Discussing and Debriefing the Activity

Have volunteers explain their work for each of the four problems, explicitly describing the arithmetic they used in finding the rates of growth. (Though this may seem routine, working through these examples will help prepare for the next activity, *Formulating the Rate,* in which students will generalize and formalize the process of finding a rate of change from points on a graph.)

Questions 1 and 2

Students found the answer to Question 1a in *How Many More People?*, but they may have worked from a table then, and here the explanation involves the coordinates of the two points. Emphasize that the answer represents the average, and does not tell us about the specific growth from year to year.

For Question 1b, students should realize that if the growth rate were constant, the graph would be a line segment connecting the two given points.

The discussion of Question 2 can proceed similarly to that of Question 1a.

Question 3: Using Different Points

After the presentation on Question 3, ask, **Did anyone use different points to calculate the average rate?** Students are likely to have selected different pairs of points to use in their computation. Bring out that the result is the same for any pair of points, and connect this observation to the fact that the points of the graph lie on a straight line.

Question 4: Just a Graph

Question 4 is the first example in the unit in which the graph is not in the context of a concrete situation. Therefore this problem provides a crucial link to the formulation of the general concept of slope. As with Question 3, bring out that students get the same result no matter which pair of points they use.

Key Question

Did anyone use different points calculate the average rate?

All in a Row

Intent

This section focuses on the study of slope.

Mathematics

The central unit problem deals with rate of change in a situation with exponential growth. In preparation for looking at the exponential functions, it is necessary that students understand average rate of change and, therefore, slope. In these activities, students develop a formula for slope and connect slope with graphs, tables, linear equations, and average rate of change in various situations. They learn to find the equation of a line, given either two points on the line or the slope and a single point.

The POW in this section engages students in exploring arithmetic sequences.

Progression

Students develop a formula for slope in *Formulating the Rate* and connect this with tables and situations in *Rates, Graphs, Slopes, and Equations.* They then look at the slopes of parallel lines (*More About Tyler's Friends*), negative slopes, and the slopes of horizontal and vertical lines (*Wake Up!*).

In *California, Here I Come!* and *Points, Slopes, and Equations*, they develop a procedure for finding the equation of a line, given either the slope and a single point or a pair of points on the line.

The Why of the Line uses geometry to explain the fact that a straight line has a constant slope, and then *Return of the Rescue* introduces the next section, where nonlinear functions necessitate consideration of an instantaneous rate of change.

Formulating the Rate

Rates, Graphs, Slopes, and Equations

More About Tyler's Friends

Wake Up!

California, Here I Come!

POW 8: Planning the Platforms

Points, Slopes, and Equations

The Why of the Line

Return of the Rescue

Formulating the Rate

Intent

In this activity, students develop the formula for slope.

Mathematics

This activity asks students to take the generalization of the concept of rate one step further, by having them express the rate of change of a function whose graph is a straight line in terms of variables representing the coordinates of two points on the line.

Progression

Students find a general expression for the rate of change. The discussion formally introduces the concept of slope and emphasizes that slope is a constant on a straight line.

Approximate Time

25 to 30 minutes for activity

15 minutes for discussion

Classroom Organization

Small groups, followed by whole-class discussion

Doing the Activity

Because the idea in this activity is so fundamental, it's important that all groups have enough time to develop the formula on their own. Some groups may find the activity trivial, and finish quite quickly. You can suggest that they work on tonight's homework assignment while other groups complete the activity.

Discussing and Debriefing the Activity

Let one or two students present and explain the expression that their groups developed. You should be able to develop a class consensus that the appropriate expression is

$$\frac{y_2 - y_1}{x_2 - x_1}$$

Post this expression and tell students that this ratio is called the **slope** of the straight line. Be sure students see that this formula represents exactly what they did in the more concrete examples of *If Looks Don't Matter, What Does?*

Tell students that the letter *m* is often used to represent slope.

Slope at Last!

Ask the class, What does slope represent? Some possible replies are:

- *The slope of a line is the amount of rise or fall of the line when you move one unit to the right.*
- *The slope is the amount that y changes per unit change in x.*
- *The slope is the rate at which y is changing as you move to the right along the line.*

(*Note*: Students will deal with the case of negative slope in the discussion of *Wake Up!*)

Also ask, How could you describe the computation of slope in words? For instance, "slope is the ratio of the difference in the *y*-coordinates of two points on the line to the difference in the *x*-coordinates of the same points."

Slope Is Constant along a Line

Ask, Does it matter which two points you use in computing the slope? As needed, refer students to their results on Questions 3 and 4 of *If Looks Don't Matter, What Does?*, in which they saw that the rate was the same no matter which pair of points was used.

Ask, Can you use the expression $\dfrac{y_2 - y_1}{x_1 - x_2}$ to compute the slope?

The question may be clearer if you take a specific line and specific points to illustrate what you mean. Extend the example to bring out that while it doesn't matter which point is "point 1" and which is "point 2," it is necessary to be consistent in the order in which they're used. The subscript notation is one way to emphasize that. Students should see that the expression $\dfrac{y_2 - y_1}{x_1 - x_2}$ is incorrect, but

that they will get the correct slope if they use either $\dfrac{y_1 - y_2}{x_1 - x_2}$ or $\dfrac{y_2 - y_1}{x_2 - x_1}$.

Rates as Slopes

Have the class return to Questions 1 through 4 of *If Looks Don't Matter, What Does?* and find the slopes of the graphs in each of those problems, using the expression they've now developed.

Point out that when a straight-line graph comes from a real-world problem, the slope has units related to the problem. For instance, in Question 1, the numerical value of the slope of the line through the two points is 20,200,000, but this number refers to people per year. In Question 2, the slope is approximately 11, and this number refers to miles per day. In Question 3, the slope is in dollars per week. Bring out that in contrast to these examples, the slope for Question 4 is a "pure number."

Are Rates Constant Along Curves?

To emphasize that constancy of slope is a property specific to straight lines, you might have students each pick two points on the graph of the equation $y = x^2$ and compute the ratio $\frac{y_2 - y_1}{x_2 - x_1}$. Then have two or three students state their results to show that the ratios are not all the same.

Ask the class what these ratios represent in each case. They should see that for any given pair of points, the ratio is the slope of the *line* that goes through those two points.

Key Questions

What does slope represent?

How could you describe the computation of slope in words?

Does it matter which two points you use in computing the slope?

Can you use the expression $\frac{y_2 - y_1}{x_1 - x_2}$ to compute the slope?

Does the slope have units?

Rates, Graphs, Slopes, and Equations

Intent

This activity gives students further experience working with ideas related to slope.

Mathematics

In this activity, students work with slope in relation to graphs, tables, and real-life situations, and they identify the slope and y-intercept in the algebraic form of a linear equation.

Progression

Students work with slope in several situations. They discuss briefly how rates of change are related to the unit problem.

For the next several activities, students will explore more details about slope and linear equations. The more general discussion of rates will resume with *To the Rescue.*

Approximate Time

30 minutes for activity (at home or in class)

10 to 15 minutes for discussion

Classroom Organization

Individuals, followed by whole-class discussion

Doing the Activity

Students work on this activity independently.

Discussing and Debriefing the Activity

You might want to assign presentation of each problem to a particular group, give them a few minutes to prepare transparencies of their graphs, and then have students make their presentations.

Questions 1 and 2

As students present each of Questions 1 and 2, elicit that the slope is numerically the same as the rate, and that this number appears in the equation as the coefficient of t.

You might also bring out, as in the discussion of *Formulating the Rate*, that in a graph that comes from a real-world context, it makes sense to assign units to the slope (feet per minute for Question 1, cubic feet per hour for Question 2).

Next, review the concept of a *y*-intercept by asking, **What's the value of the function from Question 2 when** $t = 0$**?** Then ask, **What do we call the point where the graph crosses the vertical axis?**

Bring out that both pieces of information provided in Question 2—the initial amount in the reservoir and the rate at which the reservoir is filling—are readily visible in the equation, which should look something like $v = 200,000 + 7000t$, and that the numbers 7000 and 200,000 are the slope and *y*-intercept of the graph.

Then return to Question 1 and ask, **Why don't you see the *y*-intercept in the equation for Question 1?** (which probably is in the form $d = 500t$). Bring out that the *y*-intercept in this problem is 0. You can suggest that students might want to rewrite the equation as $d = 0 + 500t$.

The Standard Slope-Intercept Form

Tell students that this form of a linear function, in which the slope and *y*-intercept are so plainly visible, is called the *slope-intercept form*.

You might mention that many books consistently use the letter *b* for the *y*-intercept (and as noted earlier, *m* is the standard letter for slope), so this form of the equation is often called the "*mx + b*" form. (If you didn't mention previously that *m* is traditionally used for slope, you can do so now.)

Questions 3 and 4

You may want to begin the discussions for Questions 3 and 4 by letting several students explain how they got the slope, in order to bring out that the result does not depend on the choice of points. Note that in these problems, students find the slope before writing the equation for the line, whereas in Questions 1 and 2, they write the equation first, then can use that to identify the slope.

Finding equations for Questions 3 and 4 may be more difficult than it was for Questions 1 and 2, because these problems are provided without context and because the problems do not explicitly give the value of the function at $x = 0$.

Elicit as many approaches as students can provide. Whatever method they use to get the equation, you can have them check that the points in the table do, in fact, fit the equation.

For Question 4, students will probably get the equation by reading off the *y*-intercept from the graph and then using something equivalent to the slope-intercept method. Allow students to present other approaches if they have them.

Reminder of the Unit Problem

This is a good time to remind students of the main unit problem. Ask, What is the connection between the unit problem and rates of change? Bring out that the unit problem involves understanding the rate at which the population has been growing, in order to predict how it will grow in the future.

Ask, Can't you just apply the idea of slope to the data from the unit problem? As needed, bring out that the definition of slope refers to graphs that form straight lines, and that the population data do not lie on a straight line. Tell students that they will learn about a generalization of slope that can be used for graphs that are not straight lines.

Key Questions

What's the value of the function from Question 2 when $t = 0$?

What do we call the point where the graph crosses the vertical axis?

Why don't you see the *y*-intercept in the equation for Question 1?

What is the connection between the unit problem and rates of change?

Can't you just apply the idea of slope to the data from the unit problem?

Supplemental Activities

Solving for Slope (reinforcement or extension) asks students how to find the slope of a linear equation when it is in the form $Ax + By = C$.

Slope and Slant (extension) and *Predicting Parallels* (extension) look at the connection between slope and angle of inclination.

More About Tyler's Friends

Intent

In this activity, students find the connection between slope and parallel lines.

Mathematics

Students observe that lines with the same slope are parallel.

Progression

This activity returns to the situation from *Story Sketches*, in which Tyler's friends were saving part of their allowances. Students graph the savings, find the slope for each graph, and make observations.

Approximate Time

30 minutes for activity

10 minutes for discussion

Classroom Organization

Small groups, followed by whole-class discussion

Doing the Activity

Students can start on this activity without any introduction. Choose a group that finishes early to prepare a transparency of the two graphs.

Note: If students have their earlier work (from Question 2 of *Story Sketches*), they don't need to redo the graphs, and can go straight to Question 2 of this activity.

Discussing and Debriefing the Activity

Have the group you selected display its two graphs and explain its answers for Question 2. They should point out that Robin's and Max's graphs have the same slope and that this slope is the same as the weekly allowance each gets.

You may want to elicit that the slope for this activity represents dollars per week (or cents per week, if students changed units).

Question 3: Slope and Parallel Lines

Have a volunteer discuss Question 3. Students will likely have observed that the lines do not meet. (Use the word *parallel*.) Be sure to get an explanation for why the lines are parallel. For instance, a student might point out that Max starts out with $2.00 more than Robin, and keeps that "lead" because they are saving the same amount of their allowance each week. Since Max always has more money, the graphs never intersect.

Another approach is more geometric, and builds on the idea that having the same numerical slope means that the lines are "equally steep." You can help students relate this informally to the angle at which they meet the *x*-axis. (The concept of *angle of inclination* is introduced in the supplemental activity *Slope and Slant*.)

Ask the class, How would you generalize the observations from this activity? They may be ready to state the general principle that lines with the same slope are parallel, but if not, you can leave the question open. The activities over the next few days will solidify this principle.

Question 4

Ask for students to give the formulas for each of the childrens' savings in terms of *t*, and then have a volunteer explain the connection to Question 2. As in the discussion of the previous activity, students should see that the coefficient of *t* is the same as the slope of the line. Here, that number represents the rate at which their savings are growing.

Also ask about the meaning of the constant terms of the formulas. Bring out that these values are Robin's and Max's starting values, and correspond to the *y*-intercepts of the two graphs.

You might encourage students to use this growth metaphor even when dealing with a context-free linear function such as Question 4 of *Rates, Graphs, Slopes, and Equations*. That is, the *y*-intercept can be thought as a starting value and the slope as a rate of growth.

Key Question

How would you generalize the observations from this activity?

Wake Up!

Intent

In this activity, students learn the meaning of negative slope.

Mathematics

Students draw graphs and write equations for situations involving negative slope. The subsequent discussion considers the special cases of slope for horizontal and vertical lines.

Progression

Students work on the activity individually and then discuss their results as a class.

Approximate Time

35 minutes for activity (at home or in class)

30 to 40 minutes for discussion

Classroom Organization

Individuals, followed by whole-class discussion

Doing the Activity

Students work on this activity independently.

Discussing and Debriefing the Activity

Let one or two students present their results. They should get a graph like this:

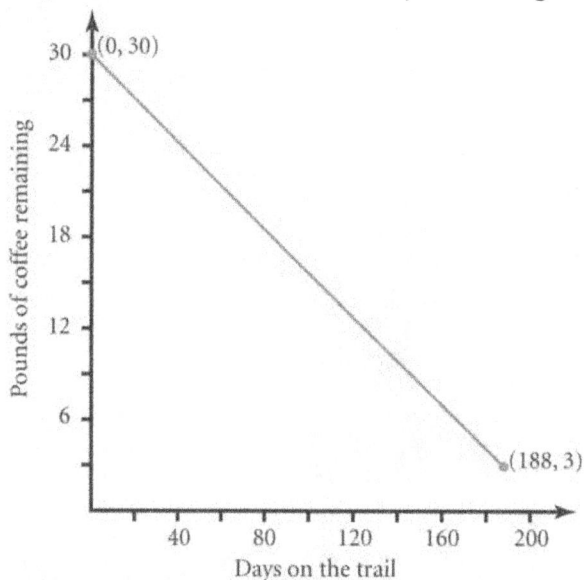

Questions 2 and 3

You can then have other students discuss Questions 2 and 3.

Other than rounding considerations, Question 2 is merely arithmetic (although some students might convert the answer from approximately 0.14 pounds to approximately 2.3 ounces).

On Question 3, using d to represent the number of days gone by, they might represent the amount remaining (in pounds) as either $30 - 0.14d$ or as $-0.14d + 30$. The first is probably a more natural representation of the situation, since it shows the initial amount and subtracts the amount consumed, but some students may be developing the habit of starting with the variable term.

Ask, **What is the connection between Question 2 and Question 3?** By now, students should not be surprised that the average rate shows up as the coefficient of the independent variable. (You can hold off on discussing the issue of sign until Question 4.)

Also ask what the number 30 represents in terms of the graph and the situation. Elicit that it is both the y-intercept and the starting value for the function.

Question 4

Let a volunteer present Question 4. Don't be surprised if there is some confusion about sign, since this is the first example students have seen involving negative slope.

If the presenter gives the slope as a positive value, you can ask the class, **How might you distinguish between the slope of a line going up and the slope of a line going down?** Bring out that it's a natural approach to assign a negative slope to a line that goes down (as it goes to the right), and that this is consistent with both the coefficient of d in the formula (from Question 2) and with the formal definition of slope as a ratio.

You may want to have the presenter identify the coordinates of the two points used in computing the slope, which were probably (0, 30) and (188, 3). Review the principle that the labeling as "point 1" and "point 2" needs to be consistent. No matter how the two points are labeled, either the numerator or the denominator of the slope fraction (but not both) will come out negative.

If the class wrote the expression for "amount remaining" as 30 – 0.14*d*, you may want to suggest that they think of this as 30 + (–0.14)*d* instead, so students are better able to see that the expression involves a negative coefficient.

Question 5

The different parts of Question 5 are quite similar to Questions 1 through 4, although they are posed in a different sequence. Again, bring out that the graph is going down (as it goes to the right), and that this fact indicates a negative slope.

After students have completed the discussion of Question 5, point out that Question 5 had them do the various tasks in a different order from the order used in the coffee problem, and ask, **In what order do you like to draw the graph, find the slope, find the average rate, and find the equation?** They may see that there is more than one reasonable sequence, but might agree that certain parts should come before others.

Horizontal and Vertical Lines

Follow the discussion of negative slopes by asking, **What is the slope of a horizontal line? What does its equation look like?** You might give students a specific example, such as the horizontal line through the points (4, 3) and (7, 3). Students should be able to find that the slope is zero by applying the definition of slope to the coordinates of the given points. (If they have trouble getting an equation, you can ask for some other points on this line.)

Specifically ask, **Is the coefficient of *x* equal to the slope in this case?** Bring out that the equation $y = 3$ can be written as $y = 0 \cdot x + 3$, so that for a horizontal line, the coefficient of *x* is 0—the same as the slope of the line.

Ask, **Does a slope of 0 for horizontal lines make sense in terms of rate of change?** They should see that horizontal lines indicate no change in *y*, so the formal definition here is consistent with the intuitive idea.

Then ask, **What is the slope of a vertical line?** Again, a specific example should help, and students should see that the definition of slope leads to an expression with 0 in the denominator. Tell the class that because of this, we do not define slope for vertical lines. Connect this with the idea that when the equation is in the form $y = mx + b$, the coefficient of *x* is equal to the slope. Bring out that for the equation of a vertical line, such as $x = 5$, one cannot solve for *y* in terms of *x* because *y* does not appear in the equation.

Ask, **How does not defining slope for vertical lines fit with the idea of slope as a rate of change?** Bring out that for a vertical line, there is no change in *x*. When we think of a function as representing change, or growth, the horizontal axis usually represents time. Thus, if time has "stood still," then it makes sense that we would be unable to measure the rate of change.

Key Questions

What is the connection between Question 2 and Question 3?

How might you distinguish between the slope of a line going up and the slope of a line going down?

In what order do you like to draw the graph, find the slope, find the average rate, and find the equation?

What is the slope of a horizontal line? What does its equation look like?

Is the coefficient of x equal to the slope in this case?

Does a slope of 0 for horizontal lines make sense in terms of rate of change?

What is the slope of a vertical line?

How does not defining slope for vertical lines fit with the idea of slope as a rate of change?

California, Here I Come!

Intent

The main purpose of this problem is to suggest to students a systematic way to use two data points to calculate slope, then use the slope and one of those data points to develop an equation for a straight line.

Mathematics

In this activity, students develop the equation of a line from two data points as they use a linear function to model population growth. They compare a value calculated using the function to another actual data point, and observe that a linear model does not fit population growth well.

Progression

Students work on the activity individually and then discuss their results as a class.

Approximate Time

25 to 30 minutes for activity (at home or in class)

20 to 30 minutes for discussion

Classroom Organization

Individuals, followed by whole-class discussion

Doing the Activity

Students work on this activity independently.

Discussing and Debriefing the Activity

Let individual students describe how they got the answers to Question 1 and Questions 2a, b, and c.

The answers are:

- Question 1: 28,740
- Question 2a: 1,529,600
- Question 2b: 2,966,600
- Question 2c: 4,403,600

It's important that they give details of the arithmetic for the parts of Question 2, because this arithmetic pattern is the key to Question 3. For instance, on Question 2a, a student might say that 1900 is 40 years after 1860, so the population would be 40 · 28,740 people more than the population in 1860. Therefore, the population in 1900 would have been 380,000 + 40 · 28,740. (Students might instead work from the year 1850—that's fine also.)

Before going on to Question 3, ask, **What kind of graph would all these points give?** (You need not have them actually plot the points.) Students should realize that they would get a straight-line graph.

Also ask, **What would be the slope of this line?** They should see that the slope is the same number as the average population growth that they found in Question 1.

Question 3

Let a volunteer describe how he or she found the formula for Question 3. The key step is recognizing that the number of years gone by from the year 1860 to the year X is $X - 1860$. (If students used 1850 as their base in Question 2, they will probably do so again here.) Thus, the amount of growth from 1860 to year X is $28,740(X - 1860)$. They also need to remember to add this growth expression to the population in 1860, giving the equation

$$P = 380,000 + 28,740(X - 1860)$$

[or $92,600 + 28,740(X - 1850)$] for the population in the year X.

Note: Some students may realize that they want to find the time elapsed, but mistakenly label this as X. If they make this error, they will probably come up with an equation like $P = 380,000 + 28,740X$. Bring out that X is defined in the problem as the actual year, so that the time elapsed is represented by the expression $X - 1860$.

Assure students that it makes sense and is correct to leave the formula for the population in the form $P = 380,000 + 28,740(X - 1860)$. But also ask, **What would be the coefficient of X if you multiplied this expression out?** They should see that they would get an expression of the form "28,740X + something," in which the coefficient of X is the average population growth.

Tell students that a formula like this is called a *linear model* for population growth. Ask, **Why do we call this a linear model for population growth?** Be sure the class sees both the algebraic and geometric meanings:

- The expression 28,740(X − 1860) + 380,000 is a linear expression.
- The graph based on this expression is a straight line.

Tell students that they will see a nonlinear model later in the unit.

As with other problems, bring out that the following numbers are all equal:

- the slope of the straight line graph
- the average annual rate of population growth
- the coefficient of X in the population formula

Question 4

Finally, see what explanations students offer for the discrepancy found in Question 4. (The official census population in 2000 was 33,871,648.) At this point, it is enough for them to say that a line doesn't fit the population values well, or that the population is growing by more and more each year.

Note: It would appear from this discussion that the population growth rate was faster from 1860 to 2000 than during the 1850's. However, students will see toward the end of the unit (*California and Exponents*) that, while the linear growth model analyzed in *California, Here I Come!* significantly *underestimates* the 2000 population, an exponential growth model vastly *overestimates* the population. (It gives an estimate of about 146 trillion.) Thus, it perhaps makes more sense to say that the population growth slowed down after the 1850's than to say that it speeded up. In the discussion of *Blue Book*, students will review the concepts of *relative* growth rate and *absolute* growth rate and see that when the relative growth rate is constant, an exponential model is a natural fit for population growth.

Optional: The Politics of Census

If you like, this activity can be used as a catalyst for a tangential discussion (no pun intended) on the issue of who got counted when the census was taken.

There is still controversy today about the accuracy of the census, with some people claiming that inner-city or rural populations are undercounted.

Key Questions

What kind of graph would all these points give?

What would be the slope of this line?

What would be the coefficient of X if you multiplied this expression out?

Why do we call this a linear model for population growth?

POW 8: Planning the Platforms

Intent

Students solve a complex problem requiring two general formulas in order to answer the questions in the POW.

Mathematics

This POW provides a context for the exploration of arithmetic sequences and series.

Progression

Give students about a week to work on the POW. Assign several students to make presentations of their solutions. The discussion introduces the terminology of arithmetic sequences.

Approximate Time

20 minutes for introduction

1 to 3 hours for the activity (at home)

25 minutes for presentations and discussion

Classroom Organization

Individuals, followed by whole-class discussion

Doing the Activity

Have students read the problem and then examine a specific case. For example, have students consider a structure with 4 platforms in which the first platform is 14 inches high and each platform is 10 inches taller than the one next to it. You might have them sketch the situation, resulting in a diagram something like this:

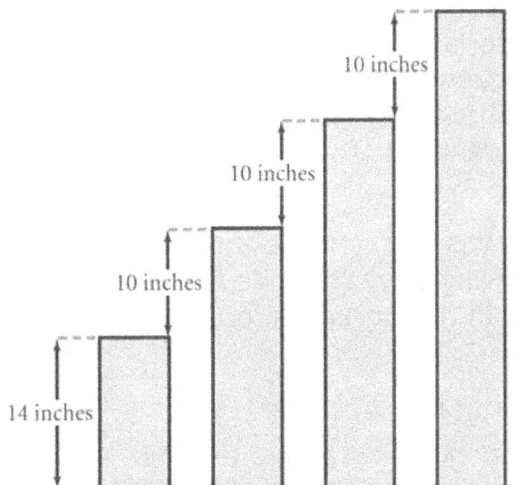

According to the problem, Camilla needs to know two things:

- the height of the tallest platform
- the total length of material needed

It is important that students see that "the total length of material needed" can be found by adding the heights of all the platforms. Thus, in this case, the platforms are 14 inches, 24 inches, 34 inches, and 44 inches tall, so the highest platform is 44 inches tall and the total length of material needed is $14 + 24 + 34 + 44 = 116$ inches.

The Task of the POW

After discussion of this example, emphasize to the class that they are to come up with two general formulas to answer Camilla's questions. Suggest to students that they introduce variables to represent the results of the three decisions Kevin must make.

Give students about a week to work on this POW. You may want to check students' problem statements after a couple of days to be sure that they have understood the task.

The Mathematical Task

This is a good POW for having students write problem statements that focus on the "pure mathematics" of the problem, independent of the contextual "story."

As students will see, the heights of the platform form an **arithmetic sequence.** (We recommend that this term be introduced following the discussion of the POW on the day that it is due.) Thus, a formula for the height of the tallest platform is equivalent to a formula for the nth term of an arithmetic sequence, and a formula for the total length of material is equivalent to a formula for the sum of the first n terms of an arithmetic sequence. Students will be expressing these quantities in terms of the initial term, the difference, and n. (In the example just discussed, the initial term is 14, the difference is 10, and n is 4.)

On the day before the POW is due, choose three students to make POW presentations on the following day, and give them overhead transparencies and pens to take home to use in their preparations.

Discussing and Debriefing the Activity

Let the selected students make their presentations. They will likely use variables to represent the number of platforms, the height of the first platform, and the difference in height between adjacent platforms. We will use *n, h,* and *d* here to represent these three pieces of information. In terms of these variables, the presenters should get the general formula

$$h + (n - 1)d$$

for the height of the tallest (*n*th) platform.

If hints are needed, ask, **What are the heights of the first few platforms in terms of the variables?** For instance, students should see that the first four platforms have heights *h, h + d, h + 2d,* and *h + 3d.* As a next hint, you might ask, **What would be the height of the 100th platform?** Get an explanation for why this is *h + 99d* rather than *h + 100d.* From that example, students should be able to get the general formula.

The Total Length of Material

For finding the total length of material needed, many students only get as far as representing this as the sum

$$h + (h + d) + (h + 2d) + \ldots + [h + (n - 1)d]$$

You might use this discussion as an opportunity to review the use of summation notation. Help students to see that this sum can be represented as

$$\sum_{k=1}^{n} \left[h + (k-1)d \right]$$

(If students prefer $\sum_{k=0}^{n-1} \left(h + kd \right)$, that's fine.)

Developing the Closed Form

Finding a way to express this sum in closed form may be too challenging for most students to do without direction. There are several common approaches for developing the closed form, all of which can be clarified through the use of specific examples. A few options are given here. Use your judgment about whether to present just one of these, give students several methods, or simply skip working on the closed form. In the last case, you might have students look at the supplemental activity *Summing the Sequences—Part I,* which concerns sums of arithmetic sequences.

The "Pairing" Method: In this method, terms are combined two at a time (first and last, second and next-to-last, and so on) to give a set of sums that are each equal to $2h + (n - 1)d$. Since there are n terms to begin with, there must be $\frac{n}{2}$ pairs (if n is even), so the sum is equal to

$$\frac{n}{2} \cdot \left[2h + (n - 1)d \right]$$

One complication of this approach is dealing with the case where n is odd.

A "Two Sums" Method: A variation of the pairing approach involves writing the sum out twice, with one sum written in reverse order. In this method, too, terms can be added in pairs. In this approach, we get n sums, each equal to $2h + (n - 1)d$, as shown:

$$
\begin{array}{ccccccc}
 & h & + & (h + d) & + \ldots + & [h + (n - 2)d] & + & [h + (n - 1)d] \\
+ & [h + (n - 1)d] & + & [h + (n - 2)d] & + \ldots + & (h + d) & + & h \\
\hline
 & [2h + (n - 1)d] & + & [2h + (n - 1)d] & + \ldots + & [2h + (n - 1)d] & + & [2h + (n - 1)d]
\end{array}
$$

Using this approach, we find that *twice* the original sum is equal to $n \cdot [2h + (n - 1)d]$. Therefore, the desired sum is half of this, which can be written

$$\frac{n \cdot \left[2h + (n - 1)d \right]}{2}$$

A Geometric Method: An approach that fits with the geometrical basis of this problem begins by placing the platforms next to each other to create a structure:

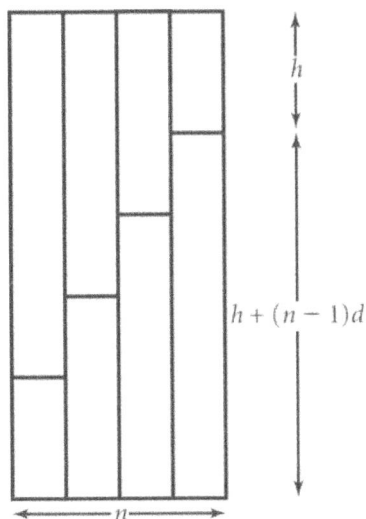

We can then imagine a second, identical structure placed upside-down on top of the first to create a rectangular figure:

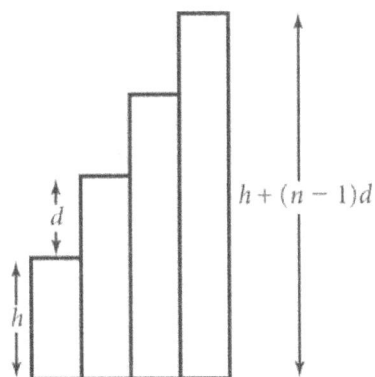

The overall height of this figure is the sum of the heights of the first and last platforms, which is again $2h + (n - 1)d$. Since the width is n (the number of platforms), the total area is $n[2h + (n - 1)d]$. As in the more numerical "two sums" method, this gives the expression $\dfrac{n\left[2h + (n-1)d\right]}{2}$ for the amount of material needed for the original structure.

Illustrate with Numerical Examples

Whatever method students use for getting a general formula, they should be encouraged to check it by means of some simple examples.

For instance, using the case mentioned in the introduction to the POW, we have $h = 14$, $d = 10$, and $n = 4$. Substituting these values into the expression $\dfrac{n\left[2h + (n-1)d\right]}{2}$ gives 116, which is the correct answer.

The Terminology of Arithmetic Sequences

Before leaving the problem, tell the class that sums in which there is a constant difference between terms (such as those occurring in this POW) are called **arithmetic sequences.** (You might point out explicitly that in this context, the word "arithmetic" is an adjective, and is pronounced "a-rith-met'-ic," with the main stress on the third syllable.) Tell them also that the first number in the sequence is called the *initial term* and that the amount added to get each successive term is called the *difference.*

Have students give you several specific numerical examples of arithmetic sequences to clarify the terminology.

Key Questions

What are the heights of the first few platforms in terms of the variables? What would be the height of the 100th platform?

Supplemental Activities

Summing the Sequences—Part I (reinforcement) deals with arithmetic sequences. In that activity, students first work with specific examples and then develop general formulas for the nth term of an arithmetic sequence and for the sum of the first n terms of an arithmetic sequence.

Summing the Sequences—Part II (extension) defines *geometric* sequences, has students work with some specific examples, and then asks them to develop general formulas for the nth term of a geometric sequence and for the sum of the first n terms of a geometric sequence. It also has them examine the limit of that sum for a specific case and then try to generalize that result.

Points, Slopes, and Equations

Intent

This activity continues students' work with slope and equations of straight lines.

Mathematics

In this activity, students write an equation of a line given the slope and one point, and then using two points. They also write equations for horizontal and vertical lines.

Progression

Students work on the activity independently and then discuss their results as a class.

Approximate Time

30 minutes for activity (at home or in class)

10 minutes for discussion

Classroom Organization

Individuals, followed by whole-class discussion

Doing the Activity

Students work on this activity independently.

Discussing and Debriefing the Activity

Part I

You can begin by having a volunteer present his or her approach to Question 1, and then ask for other methods. Here are two approaches that may come up:

- The slope is 5, so the equation should look like $y = 5x + c$. When x is 3, y is 2, so c must be -13. So the equation is $y = 5x - 13$.
- When x is 3, y is 2, and y goes up by 5 for each unit increase in x. The amount by which x has increased can be represented by the expression $x - 3$, so the increase in y is $5(x - 3)$. [For instance, if x goes up to 7, it increases by $7 - 3 = 4$, so y increases by $5(7 - 3)$.] Therefore, the value of y is $2 + 5(x - 3)$. (This approach is similar to that described for the discussion of *California, Here I Come!*)

If a student presents the equation in the form $y = 2 + 5(x - 3)$ (the second approach), bring out how clearly it shows the slope of 5 and the point (3, 2), and identify this as the *point-slope form* of a linear equation. However, you need not press students to develop a general point-slope formula here. Rather, help them see the reasoning involved in writing an equation in this form.

Once the class has found the equation (perhaps in several different ways), have them verify that the point (3, 2) really fits the equation. Also, have them find a second point that fits the equation and verify that the new point and (3, 2) do indeed lead to a slope of 5.

Questions 2 through 4

If students had trouble with Question 1, you can have them work in their groups on Questions 2, 3, and 4 now. Other than issues of sign (including the use of a negative slope in Question 3) and the use of a fractional slope in Question 4, these three problems are quite similar to Question 1.

Part II

Students will probably not have trouble finding the slope in these problems, so the questions in Part II are quite similar to those in Part I. Bring this point out in the discussion, so students realize that they can always find the equation if they are given two points.

Summary

If it hasn't already come out clearly in the discussion, bring out the connection here between the algebra of students' work and the geometric principles behind it. That is, they should see clearly that just as a line can be determined geometrically by two points or by one point and a direction, so also a linear equation can be found if one knows either two points on its graph or one point and the slope.

Part III

Use the examples in Part III to review the idea that horizontal lines have a slope of 0 and the slope is undefined for vertical lines. (These ideas were discussed following *Wake Up!*)

The Why of the Line

Intent

In this activity, students use geometry to explain why using any pair of points on a given line will lead to the same slope.

Mathematics

Until now, students have taken for granted that a line has a constant slope. In this activity, they explain this fact using ideas from geometry. The goal is for them to understand that the constant slope of lines is related to the concept of similarity.

Progression

Students work in groups to explain why the slope of a line does not depend on the choice of points, using ideas from geometry. The discussion brings out the role of similarity, reviewing principles for determining similarity and principles concerning parallel lines. The discussion also briefly introduces the next section, in which the focus turns from straight lines to curves.

Approximate Time

25 to 30 minutes for activity

10 to 15 minutes for discussion

Classroom Organization

Small groups followed by whole-class discussion, or just whole-class discussion

Materials

Transparency of *The Why of the Line* blackline master

Doing the Activity

You may need to treat this activity primarily as a whole-class discussion, perhaps with strong guidance from you, but give students some time to work on it in groups before bringing them together.

Discussing and Debriefing the Activity

Begin by having one or two students share their ideas. Students should see that the two right triangles *AEB* and *CFD* are similar. Before asking for a proof that the triangles are similar, make sure the class sees that this similarity leads to the conclusion that $\dfrac{r}{s} = \dfrac{t}{u}$. (Students may instead see the proportionality as $\dfrac{r}{t} = \dfrac{s}{u}$. If so, then they will need to do a bit of algebra to see why the two slopes are equal.)

Why Are the Triangles Similar?

Ask, **How can you be sure that the two triangles are similar?** If necessary, review the principle that if two triangles have two pairs of corresponding angles equal, then the triangles must be similar.

In this situation, both triangles have a right angle, so students begin with one pair of equal angles. Perhaps the best way to get a second pair of equal angles is to use angles *BAE* and *DCF*. Students might reason like this:

> Since $\overline{AE}$ and $\overline{CF}$ are horizontal, they are parallel. This makes the line through A, B, C, and D a transversal across parallel lines, so the corresponding angles BAE and DCF must be equal.

Emphasize that this reasoning works precisely because the line through *A* and *B* and the line through *C* and *D* is the same line; that is, it works because the four points *A, B, C,* and *D* are collinear.

You may find it helpful to make a small cutout triangle that is similar to these two triangles. You can place this with its hypotenuse along the line, and slide it along to see that angles *BAE* and *DCF* are congruent.

You also might use a diagram like the one shown here, with points *A, B, C,* and *D* placed on some non-linear graph, to point out that if the points are not collinear, then angles *BAE* and *DCF* need not be equal and right triangles *AEB* and *CFD* need not be similar.

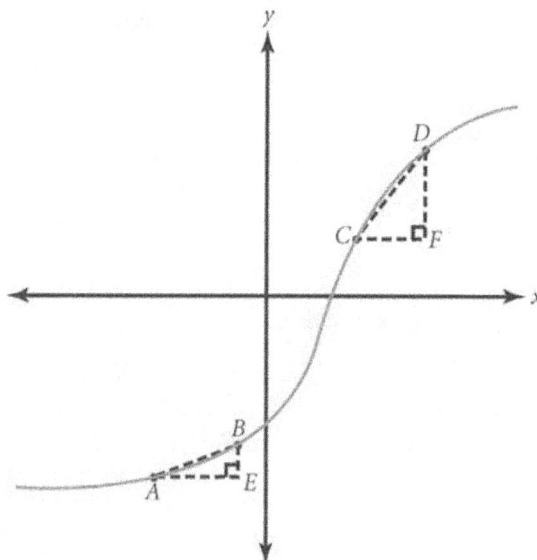

From Straight Lines to Curves

This concludes a major segment of the unit, dealing with linear equations and slope. You can begin the transition to the next phase of the unit by asking, **What is the task in the central unit problem?** Bring out that they are looking for a mathematical description of how the population has been changing so that they can predict when it might reach a certain value.

Point out that they have been working for the past week or so with linear functions, but they saw in *California, Here I Come!* that the linear model for population growth was not a very good predictor, since the rate at which population grows does not seem to be constant.

Tell them that the next activity begins the next phase of the unit, in which they will develop a generalization of slope that applies to nonlinear situations.

Key Questions

How can you be sure that the two triangles are similar?

What is the task in the central unit problem?

Supplemental Activity

The Slope's the Thing (extension) continues students' work with finding the equation of a line, given the coordinates of two points on the line.

To the Rescue

Intent

This activity concludes this portion of the unit. While still focused on *average* rate of change, the equation in the problem is not linear.

Mathematics

In this activity, students answer questions about a falling object. They see the average speed as the slope of a secant line.

Progression

Students work on this activity individually and then discuss their results as a class. The discussion introduces the term **secant line.**

Approximate Time

30 minutes for activity (at home or in class)

15 minutes for discussion

Classroom Organization

Individuals, followed by whole-class discussion

Materials

Transparency of *To the Rescue* blackline master

Doing the Activity

Students work on this activity independently.

Discussing and Debriefing the Activity

For Question 1, the idea of substituting $t = 3$ into the formula may be so simple that it eludes some students. Be sure that everybody realizes that when they are given such a formula, they just have to plug in and do the arithmetic. (We have included Question 1 explicitly to help students distinguish between the object's height and the distance it has fallen.)

Question 3 requires students to know what average speed means. After students answer the specific question, ask, **In general, how do you find average speed?** Students should state that average speed is simply the ratio

$$\frac{\text{distance traveled}}{\text{time elapsed}}$$

Speed as a Slope

Before going on to Question 4, have students graph the function $h(t)$ on a graphing calculator. Then show students a labeled graph like this one, but without the segment connecting (0, 400) to (3, 256). (This graph is provided in the *To the Rescue* blackline.)

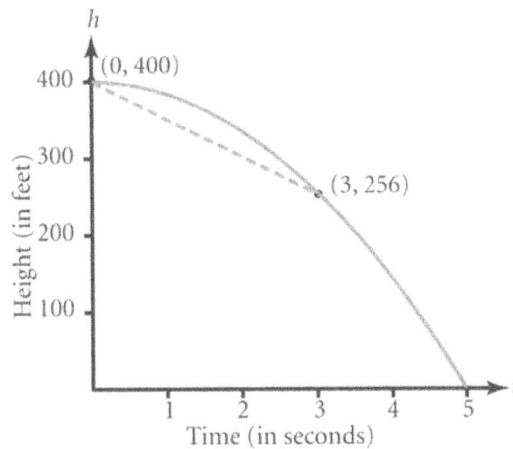

Ask, **What does the average speed mean in terms of the graph?** Then ask, **How can you represent the distance traveled and the time elapsed as lengths of vertical and horizontal line segments on the graph?**

Students should be able to articulate that except for sign, the average speed is the slope of the line segment connecting the points (0, 400) and (3, 256).

Ask students, **How can you represent this slope as a ratio using the notation $h(t)$?** They should see that the slope is equal to the expression

$$\frac{h(3) - h(0)}{3 - 0}$$

in which the numerator is the distance traveled and the denominator is the amount of time elapsed.

Don't let students get too distracted here by the issue of sign. You can comment that the slope is negative because the height is decreasing, but that we define speed to always be non-negative. You might identify it as the absolute value of the slope.

Introduce the term **secant line** to describe the line (or line segment) connecting two points on a graph.

Question 4

Have a volunteer give and explain the answer to Question 4. Students will use the fact that the fall takes 5 seconds in the next activity, *The Instant of Impact*.

Key Questions

In general, how do you find average speed?

What does the average speed mean in terms of the graph?

How can you represent the distance traveled and the time elapsed as lengths of vertical and horizontal line segments on the graph?

How can you represent this slope as a ratio using the notation $h(t)$?

Beyond Linearity

Intent

In this section, students explore the meaning of instantaneous rate of change.

Mathematics

While the central unit problem obviously involves rate of change, the nonlinear nature of the data makes it appear that the concept of slope may be of little value in this problem. In *Beyond Linearity*, students expand the idea of slope to nonlinear functions as they explore instantaneous rate of change. Consideration of average rate of change over very small intervals leads to the slope of the tangent and to the derivative.

Progression

Four activities provide a solid introduction to the concept of instantaneous rate of change: *The Instant of Impact*, *Doctor's Orders*, *Photo Finish*, and *Speed and Slope*. *ZOOOOOOOOM* and *Zooming Free-for-All* help students connect this idea of instantaneous rate of change with the slope of the tangent.

Students consider instantaneous rate of change in a number of contexts through *The Growth of the Oil Slick* and *Speeds, Rates, and Derivatives*. The latter activity introduces the derivative.

On a Tangent illustrates that the derivative can be effectively approximated by the slope of a secant line that passes through the function at two points very close together.

The Instant of Impact
Doctor's Orders
Photo Finish
Speed and Slope
ZOOOOOOOOM
The Growth of the Oil Slick
Speeds, Rates, and Derivatives
Zooming Free-for-All
On a Tangent
POW 9: Around King Arthur's Table
What's It All About?

The Instant of Impact

Intent
This activity introduces instantaneous speed.

Mathematics
In this activity, students look at the average speed of a falling object for increasingly smaller increments of time just before impact, to understand speed at an instant.

Progression
The Instant of Impact continues consideration of the falling bundle from *To the Rescue*.

Approximate Time
25 minutes for activity

10 minutes for discussion

Classroom Organization
Small groups, followed by whole-class discussion

Doing the Activity
Introduce this activity by reviewing that the bundle is 256 feet above the ground at $t = 3$ (Question 1 of *To the Rescue*) and that it hits the ground at $t = 5$ (Question 4 of *To the Rescue*). Students can then use these facts in Question 1 of *The Instant of Impact*.

Tell students to keep careful track of their computations. Questions 1 through 4 are very similar to *To the Rescue*, but Question 5 is a conceptual leap. If a group seems to have a good grasp of the idea of instantaneous speed, have them prepare a presentation.

Discussing and Debriefing the Activity
Begin by having students present Questions 1a and 1b. These questions are similar to Questions 2 and 3 of *To the Rescue* and again should be interpreted in terms of the graph. Students should see that the segment connecting (3, 256) to (5, 0) is steeper than the segment connecting (0, 400) to (3, 256), which means that the supply bundle is falling faster near the end of its fall.

Then let other students present their results for Questions 2 through 4, which lay the groundwork for Question 5. Bring out through these problems that the y-values are found simply by substituting the given value of t into the equation $h(t)$.

As previously, have students identify the representation of the rate as a ratio. For example, in Question 3, they might represent the average rate of speed using the expression $\dfrac{h(4.5)-h(5)}{4.5-5}$ (although they will then need to adjust the sign).

At this time, focus discussion on the numerical details and omit reference to the graph. Graphs will be a focus in *ZOOOOOOOOM* and *Zooming Free-for-All* (in which the graph of this falling-bundle situation will be examined).

Question 5

If you had a group prepare a presentation for Question 5, begin with that presentation.

In any case, ask students, **What does 'the speed at the exact moment' mean?** Guide students to articulate the idea of considering smaller and smaller time intervals, to the point at which there is very little change. For instance, in considering the last thousandth of a second rather than the last hundredth of a second, the average speed changes only slightly, going from 159.84 feet per second (for the last hundredth of a second) to 159.984 feet per second (for the last thousandth of a second). Students should see that as the interval gets smaller and smaller, the speed computed gets closer and closer to 160 feet per second. (If students consider small-enough time intervals, they will find that their calculators give exactly 160 feet per second as the result. Be sure they realize that this is a round-off error and that the speed even during the last millionth of a second is actually slightly less than 160 feet per second.)

Once this general approach has been clarified intuitively, introduce the term *instantaneous speed* for the speed at an exact instant.

The Bundle Survives

Be sure to return to the implied question of the activity—whether the bundle will survive the impact of hitting the ground. Because the speed at that moment appears to be 160 feet per second (or very close to that), the bundle does withstand the impact. (The problem states that the bundle can withstand an impact at speeds up to 165 feet per second.)

Key Question

What does 'the speed at the exact moment' mean?

Doctor's Orders

Intent

In this activity, students continue to work with the concept of instantaneous speed.

Mathematics

This activity provides additional practice with looking at average speeds during increasingly small increments of time before the impact of a falling body, explored in the previous activity.

Progression

This activity provides additional practice with the concepts of *The Instant of Impact*. Students will benefit from this review before working on the next activity, *Photo Finish.*

Approximate Time

25 minutes for activity (at home or in class)

10 minutes for discussion

Classroom Organization

Individuals, followed by whole-class discussion

Doing the Activity

This activity presents another situation involving instantaneous speed. Students should be able to do this activity independently.

Discussing and Debriefing the Activity

Have students from different groups discuss how they approached each part of the problem.

Note: Students do not necessarily have to calculate instantaneous speed in order to answer Question 5. For example, they might see that Clayton's average speed during the last tenth of a second is already more than 60 feet per second, so his speed when he hits the water will certainly be more than that.

Photo Finish

Intent

In this activity, students continue to work with the concept of instantaneous speed.

Mathematics

Students continue to approximate instantaneous speed by looking at average speed for small increments of time, but they are given less guidance than in previous activities. Students should be becoming comfortable with this approach by now.

Progression

Students work in groups to determine the instantaneous speed of a runner at several points in time, given a formula for the distance run as a function of time.

Approximate Time

25 to 30 minutes for activity

15 minutes for discussion

Classroom Organization

Small groups, followed by whole-class discussion

Doing the Activity

This activity continues the work with instantaneous rates.

As groups finish Question 2, assign a group to prepare a presentation on it. Other groups can continue with work on Question 3a. Question 3b is an additional challenge for groups that are ready for it—if no groups get to this problem, you can skip it.

Discussing and Debriefing the Activity

Have a student present the answer to Question 1 and describe how he or she got the answer. If the student used a graph to get an estimate of 50 for t, ask how to verify that this is the exact value. If needed, review the idea of simply substituting 50 for t in the formula. (This is a good opportunity to talk about the limitations of calculator graphing.)

Use the discussion of Question 2 as a way to assess how well students understand the idea of instantaneous speed. There's no need to debate "the best" approximation of Speedy's final speed—the point is whether they realize that they want to find her speed during a very short interval at the end of the race.

For example, students might look at the last tenth of a second, that is, the interval from $t = 49.9$ to $t = 50$. As needed, bring out that they can find out how far Speedy had run as of $t = 49.9$ by substituting into the formula for $m(t)$. They should find that $m(49.9) = 398.701$. Thus, Speedy ran the remaining 1.299 meters (out of the total of 400 meters) during the final tenth of a second. This gives an average speed of 12.99 meters per second. Speedy's speed at exactly $t = 50$ is likely to be very close to this. If students look at smaller intervals, they will probably conclude that the exact speed is 13 meters per second (which is the correct value).

Question 3

If any groups have results to share from Question 3a, let them do so. If not, you can either move on or have groups work on this now.

If students do share speeds for other instants, you can also move on to discuss Question 3b. Students should realize that if Speedy was going slower than 10 meters per second for some instants and faster than this for others, then there should have been an instant when she was going *exactly* 10 meters per second. (You can use the idea of an auto speedometer to help illustrate this. Students will see, for instance, that a car can't accelerate from 20 mph to 60 mph without having the speedometer dial go through every speed in between.)

To actually find the value of t when the speed was 10 meters per second, students might use guess-and-check. (*Note:* Because the function in this problem is quadratic, a proportionality argument will also yield the right answer. See the subsection "Caution: A Special Property of Quadratic Functions" later in this section.)

Exact Values

Students may raise the question of how to determine, or be certain of, the exact value of the instantaneous speed.

In the situations considered in *The Instant of Impact, Doctor's Orders,* and *Photo Finish,* students can tell from the shape of the graph that the speed is increasing. This enables them to determine for sure that the exact value is within certain bounds.

For instance, in *Photo Finish,* if Speedy continued beyond the finish line according to the same function, her average speed between $t = 50$ and $t = 50.1$ would be 13.01 meters per second. Students can combine this with the fact that her average speed between $t = 49.9$ and $t = 50$ is 12.99 meters per second to conclude that her instantaneous speed at $t = 50$ is between 12.99 and 13.01 meters per second.

Caution: A Special Property of Quadratic Functions

Students may notice that the instantaneous speed at $t = 50$ seems to be exactly halfway between the average speed for the interval from $t = 49.9$ to $t = 50$ and the average speed for the interval from $t = 50$ to $t = 50.1$. Or they may notice that they seem able to get the precise instantaneous speed by using an interval symmetric about $t = 50$, such as from $t = 49.9$ to $t = 50.1$. They may have made similar observations for *The Instant of Impact* and *Doctor's Orders*.

If this comes up, tell students that this phenomenon is a special property of quadratic (or linear) functions. (Of course, this might occur merely by coincidence for particular values with other functions, but it will not hold true in general.)

Speed and Slope

Intent

In this activity, students summarize ideas about instantaneous speed.

Mathematics

Students summarize what they've learned about finding instantaneous speed, and practice applying concepts related to linear functions from earlier in the unit..

Progression

Students work on the activity individually and then discuss their results as a class.

Approximate Time

25 minutes for activity (at home or in class)

10 minutes for discussion

Classroom Organization

Individuals, followed by whole-class discussion

Doing the Activity

Students work on this activity independently.

Discussing and Debriefing the Activity

Have students briefly share their ideas on Part I.

Part II

On Question 1, students will probably compute the rate (6 millimeters per hour) and get the expression $420 + 6t$.

On Question 2, students will probably begin by finding the slope, but there may be some variation in how they proceed. You might have a couple of students present different approaches.

ZOOOOOOOOM

Intent

In this activity, students use a close-up of a graph to help them understand the behavior of a function.

Mathematics

The discussion of this activity introduces the concept that the slope of a tangent line to the graph of a function represents the instantaneous rate of change.

Progression

ZOOOOOOOOM returns to the situation of the falling supply bundle from *To the Rescue*—this time, students analyze the situation graphically.

Approximate Time

40 minutes for activity

25 minutes for discussion

Classroom Organization

Small groups, followed by whole-class discussion

Doing the Activity

If students ask, tell them that they can use (50, 400) as one of the two points in Question 1a. (If they are using the trace feature of the graphing calculator, this point may not show up exactly.) Except for (50, 400), though, it is important that students find their points using the trace feature, rather than from some other approach such as a table or substitution into the formula. By focusing on the graph, students will get a better sense of the geometric meaning of slope.

Note: Because of the limitations of calculator screens, the graphs students get as they zoom in may be a bit jagged. If necessary, remind them of this, and let them know that these graphs are sufficiently close to straight for them to proceed with the activity.

Discussing and Debriefing the Activity

Let a student give the coordinates he or she used for the two points on the (apparent) line and present the computation for the slope of the line. Then ask if other groups used different points, and have one or two students from those groups present their work. Unless students misread the coordinates or made computational errors, the slopes they found should all be fairly close to 13. Try to get a consensus among students that if they kept zooming in closer and closer to (50, 400), the resulting slopes would get closer and closer to 13.

Remind the class that these are really points on a curve, and ask, **What's the term for a line connecting two points on a curve?** Review the term *secant line.* Be sure it is clear that each group is computing the slope of a secant line and that different groups may have used different secant lines.

As in the discussion of *The Growth of the Oil Slick,* you may find it useful to sketch a curve that exaggerates the curvature of the graph in order to distinguish between the curve and a secant line, like this:

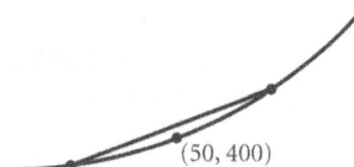

(50, 400)

Question 2

The main focus of the discussion should be on Question 2. Have volunteers share their ideas on this question. Elicit the observation that the slopes groups found all represent average speeds for short time intervals near $t = 50$. Students should see that slight changes in the time interval lead to slight changes in the resulting slope, but that small intervals including $t = 50$ all give results very similar to one another.

Ask, **What do these slopes have to do with Speedy's instantaneous speed at** $t = 50$**?** Students should articulate that they all represent approximations of that speed and that the instantaneous speed is what one gets as the time intervals get smaller and smaller.

If needed, remind students that they agreed in *Photo Finish* that Speedy's speed at the finish line was, in fact, 13 meters per second.

The Tangent Line

Ask students, **What line through (50, 400) has as its slope the exact value of the instantaneous speed?** Ask for a volunteer to sketch such a line on the graph used previously to show the secant lines.

Inform students that this line is called the **tangent line** (or just the *tangent*) to the graph at the given point. Bring out that if the two points for a secant line are very close to (50, 400), then the secant line and the tangent line nearly coincide.

Key Questions

What's the term for a line connecting two points on a curve?

What do these slopes have to do with Speedy's instantaneous speed at $t = 50$?

What line through (50, 400) has as its slope the exact value of the instantaneous speed?

The Growth of the Oil Slick

Intent

This activity asks students to apply the concept of instantaneous rate of change in a context other than speed.

Mathematics

In this activity, students find the instantaneous rate of change by looking at intervals both before and after the instant under consideration.

Progression

The Growth of the Oil Slick returns to the situation from *What a Mess!*, this time looking at the growth of the area of the oil slick, rather than the growth of the radius.

Approximate Time

30 minutes for activity (at home or in class)

10 to 15 minutes for discussion

Classroom Organization

Individuals, followed by whole-class discussion

Materials

Transparency of *The Growth of the Oil Slick* blackline master

Doing the Activity

Students work on this activity independently.

Discussing and Debriefing the Activity

Have students present their work on different parts of the activity. The concept of an instantaneous rate of change may be harder for students to visualize in this context than it was in the context of speed (for instance, in *The Instant of Impact*).

Have students sketch by hand a graph of the area (as a function of time) in order to get an intuitive sense of the instantaneous growth rate. Have them mark line segments on the graph to show the relevant values. For instance, the graph here indicates the segments needed to find the rate of change in the area for the time interval from $t = 0$ to $t = 2$. (A larger version of this graph, without the line segments, is provided in *The Growth of the Oil Slick* blackline master.

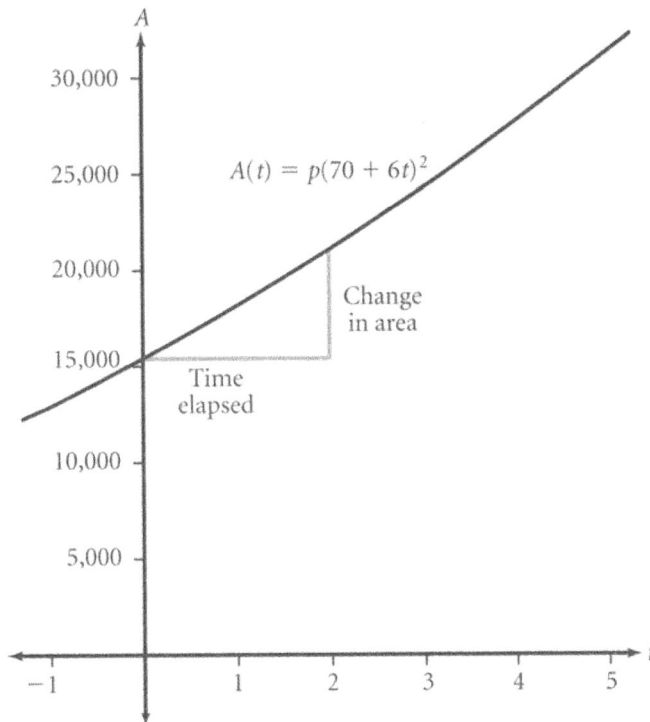

The graph shows $A(t) = p(70 + 6t)^2$, with axes labeled *A* (vertical, marked at 5,000; 10,000; 15,000; 20,000; 25,000; 30,000) and *t* (horizontal, marked at −1, 1, 2, 3, 4, 5). Annotations read "Change in area," "Time elapsed."

Note: The rate of change does not vary much within the intervals under consideration, so you may prefer to make a sketch that exaggerates the curvature of the graph. Be sure students are aware that because this is a quadratic function, its graph is not a straight line.

One new element in this activity is that it uses time intervals both before and after the instant under consideration. Bring out that in the oil slick situation, the average rates for intervals before $t = 0$ are less than the instantaneous growth rate at $t = 0$ and the average rates for intervals after $t = 0$ are greater than the instantaneous growth rate at $t = 0$. You can use a sketch of the graph to illustrate what this means in terms of secant lines. That is, because the curve is getting steeper as it goes to the right, secant lines using $(0, 4900\pi)$ as the right-hand endpoint have slopes less than the instantaneous rate, while secant lines using $(0, 4900\pi)$ as the left-hand endpoint have slopes greater than the instantaneous rate.

Reminder: This activity may suggest to students that they use a "mixed" interval—that is, an interval that begins before $t = 0$ and ends after $t = 0$. If they use an interval that is symmetric about $t = 0$, they will get the *exact* instantaneous growth rate. (For more on this issue, see the subsection "Caution: A Special Property of Quadratic Functions" at the end of the discussion of *Photo Finish*.)

Here are the answers, rounded as indicated:

- Question 1: 15,394 square meters (to the nearest square meter)

- Question 2a: 2865 square meters per hour (to the nearest square meter per hour)

- Question 2b: 2695 square meters per hour (to the nearest square meter per hour)

- Question 2c: 2611 square meters per hour (to the nearest square meter per hour)

- Question 3: 2639 square meters per hour (to the nearest square meter per hour)

π *Versus 3.14*

This is a good time to remind students that the number π is not the same as its decimal approximation, 3.14. In the context of this problem, an approximate value is sufficient, if used consistently. But be sure students are aware that the results they get have less precision than if they were using a better approximation. (You might acknowledge that even the number given by the calculator's π key is only an approximate value.)

Supplemental Activity

Potential Disaster (reinforcement) can be used if reinforcement of the principles from this activity is needed.

Speeds, Rates, and Derivatives

Intent
In this activity, students work with the concept of the derivative and connect it to familiar situations.

Mathematics
The teacher introduces this activity by defining the derivative. Students then revisit several of the situations they have encountered in this unit and find the derivative at a particular point and consider what that derivative means.

Progression
The discussion prior to the student activity introduces the concept of a derivative. In the activity, students learn to interpret derivatives as rates in real-life situations.

Approximate Time
10 to 15 minutes for introduction

30 minutes for activity (at home or in class)

10 minutes for discussion

Classroom Organization
Individuals, followed by whole-class discussion

Doing the Activity
Point out to students that the number 13 that came out of the discussion of *Photo Finish* can be described in several ways:

- It is Speedy's *instantaneous speed* at the moment she crosses the finish line.
- It is the slope of the tangent line to the graph of the function $m(t) = 0.1t^2 + 3t$ at the point (50, 400).
- It is the "end result" of finding average speeds over shorter and shorter intervals.
- It is the "end result" of finding slopes of shorter and shorter secant lines through the point (50, 400).

Tell students that in calculus, this number is called a **derivative.** More specifically, it is the *derivative of the function m(t) at the point (50, 400).* Tell them it is also called the *derivative of the function m(t) at t = 50.* Thus, the derivative of a function at a point is the same as the slope of the line that is tangent to the graph at that point.

Tell students that the "end result" process of using smaller and smaller intervals is the basis of a very technical concept called a **limit,** which they will study if they take calculus. The approach used in calculus is sketched in the supplemental activity *Speedy's Speed by Algebra.*

Discussing and Debriefing the Activity

Have students from different groups share their answers.

For Question 1a, the presenter should find the derivative of the function at $t = 3$ is –96. (If you think it's needed, review the process for getting this derivative. In particular, be sure students know how to use the formula to find the relevant numerical values.) The discussion of Question 1b should bring out that this means that the supply bundle is falling at a rate of 96 feet per second three seconds after it is dropped.

On Question 2a, the presenter should find that at $t = 1$, the oil slick is growing at approximately 2865 square meters per hour. In the discussion of Question 2b, students should see that this value represents the derivative of the function $A(t) = \pi(70 + 6t)^2$ at the point $(1, \pi \cdot 76^2)$ (or "at $t = 1$").

Question 3

The main purpose of Question 3 is to bring out that because linear functions have a constant rate of change, they have the same derivative at all points on the line, which is the same as the slope of that line. Students should also see that—except for the sign—the derivative, the slope, and the coefficient of d all represent the amount of coffee the Cazneaus drink per day.

Zooming Free-for-All

Intent

In this activity, students continue to see that for most functions a close-up of the graph at any point will look like a straight line.

Mathematics

This activity helps students to make sense of the derivative by exploring what happens when they zoom in on the graph of a function. By seeing that a close-up of the graph will appear to be a straight line, they are able to connect the derivative with the instantaneous rate of change.

Progression

Students explore zooming in on functions of their choice, observing that most selections can be made to look like a straight line. The final question asks students to think of a graph that will not appear as a straight line at some point. Discussion of this question brings out that the derivative is not defined at $x = 0$ for the absolute value function.

A brief review of the status of the unit problem follows the discussion of this activity.

Approximate Time

20 to 25 minutes for activity

5 to 10 minutes for discussion

Classroom Organization

Small groups, followed by whole-class discussion

Doing the Activity

Students explore the questions of this activity in groups, using graphing calculators.

Discussing and Debriefing the Activity

Let selected students share their observations and ideas. On Question 1, bring out that except for sign, the slope of the "apparent line" represents the speed of the falling supply bundle for the value of t at the specific point chosen (assuming that t is between 0 and 5).

On Question 2, let other students share their experiences with zooming, and use the discussion to review the terms *tangent line* and *secant line*. Bring out that the graph itself is not a straight line, but that as students zoom in, what they see closely resembles both a tangent line and a secant line.

Question 3

Students may be unable to come up with an example of a function that does not appear straight when they zoom in enough. If so, you should give them the example of the absolute value function, near the point (0, 0).

You might have them either enter and graph this function on their graphing calculators or work in groups to sketch the graph by hand. Bring out that no matter how much they zoom in on the graph near (0, 0), the resulting graph keeps its V shape, as shown here:

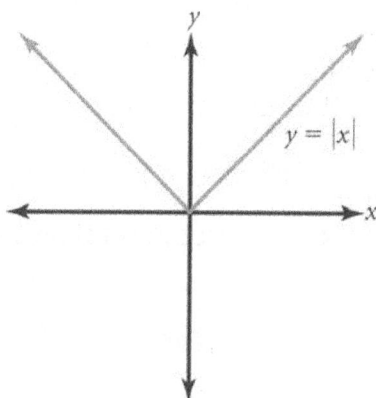

Warning: Some students may have identified a feature on their graphing calculators that calculates the derivative of a function at a given point. Many calculators will incorrectly show that the absolute value function has a derivative equal to 0 at $x = 0$—this is because the calculator calculates derivatives using numerical approximation, which can be inaccurate in special cases like this one.

Reminder of the Unit Problem

Once again, connect this work to the central unit problem. Students are studying rate of change to increase their understanding of functions, so that they can better choose a function that will help them predict when the world population will be squashed together.

On a Tangent

Intent

In this activity, students examine both secant and tangent lines and their connection to derivatives.

Mathematics

Students draw successive secant lines passing through the same point on the graph of a function, with each line more closely resembling the tangent through that point. They find that the secant through two points that are very close together approximates a tangent, and thus the slope of the secant approximates the slope of the tangent, or the derivative at that point.

Progression

Students work on this activity individually and then discuss their results as a class.

Approximate Time

30 minutes for activity (at home or in class)

10 minutes for discussion

Classroom Organization

Individuals, followed by whole-class discussion

Doing the Activity

The first problem asks students to graph a given function, using an entire sheet of graph paper. It gives them several secants to draw through a point and then asks them to draw the tangent at the same point and to find the associated derivative.

The second problem gives the graph of another function and asks students to copy the graph and draw the tangent lines at several points on the graph, estimating the derivative at each.

Discussing and Debriefing the Activity

Have a volunteer present the graph for Question 1a. Be sure that the scales and the point (2, 2) are clearly marked. Then have other students present their work on parts i through iv of Question 1c, using the graph to explain their work. It will probably be helpful in the subsequent discussion of Question 1d if you record the slopes for each case of Question 1c. (The slopes are 1, 1.5, 1.75, and 1.95.)

Next, have a volunteer present Question 1d, including an explanation of how he or she found the slope. The presentation should connect the value of this slope to the pattern of slopes found in Question 1c. Students may find it helpful to include another point or two from the graph, even closer to (2, 2), to confirm that the slopes of the secants are getting closer and closer to 2. We suggest that you describe the slope of the tangent line as the limit of the slopes of the secants even though the word *limit* has not been formally defined. (It should have been used informally in the discussion of *Speeds, Rates, and Derivatives*.)

The purpose of Question 1e is to bring out the connection between the tangent line and the derivative. You can have one or two students explain how they found the derivative. Use their presentations to elicit that the process used for finding the slope of the tangent line is exactly the same as the process for defining the derivative.

Question 2

Have different volunteers present their results for each of points *A, B,* and *C*. The derivatives students find will be fairly crude estimates, so you should focus the discussion on the general ideas rather than the numerical details. The key facts are that the derivative at *A* is negative (roughly –2), the derivative at *B* is positive (but less than 1), and the derivative at *C* is 0 (or very close to it).

Supplemental Activity

Proving the Tangent (extension) asks students to find the equation of the tangent line based on this estimated slope and then confirm that the resulting line is the tangent by showing algebraically that it meets the graph exactly once.

POW 9: Around King Arthur's Table

Intent

In this activity, students analyze and explain numerical patterns.

Mathematics

This activity has no direct connection to the mathematics of this unit. It provides students with further experience in recognizing patters and expressing those patterns as formulas, as well as in organizing and communicating their thoughts.

Progression

The POW describes a game and asks students to analyze how they might consistently win that game. Students should be given about a week to complete the POW, followed by several student presentations.

Approximate Time

15 minutes for introduction

1 to 3 hours for activity (at home)

25 to 30 minutes for presentations and discussion

Classroom Organization

Individuals, followed by several student presentations and whole-class discussion

Doing the Activity

Demonstrate the game described in the POW to make sure students understand how it works. You can have a group of students sit in a circle as you play the part of King Arthur. (Using between six and ten students works well.) After students are clear about how the game is played, they can gather more data with their groups for the remainder of the class.

On the day before the POW is due, choose three students to make presentations on the following day, and give them overhead transparencies and pens to take home to use for preparing their presentations.

Discussing and Debriefing the Activity

Ask the three students selected to make their presentations. They will probably have created an In-Out table for the problem, with headings like "Number of knights" for the *In* and "Winning chair number" for the *Out*. For instance, they might get a table like this:

Number of knights	Winning chair number
1	1
2	1
3	3
4	1
5	3
6	5
7	7
8	1
9	3
10	5
11	7
12	9
13	11
14	13
15	15
16	1

They will probably be able to identify the pattern here—the *Out* numbers are the sequence of odd numbers, but every time the *In* is a power of 2, the *Out* returns to 1. (If presenters do not bring out this pattern, you should help students to see it.)

There are two major challenges in this POW:

- Expressing this pattern using some type of formula or describing it by some procedure
- Explaining the pattern

Following are some guidelines for discussing each of these two aspects of the POW.

Expressing the Pattern as a Formula

One way of expressing the rule for this pattern is to say "write n as a power of 2 plus some more" (where n is the number of knights). In other words, write n using an formula like $n = 2^r + k$, where k is a number satisfying the condition $0 \leq k < 2^r$. In terms of this expression for n, students should be able to figure out that the winning chair number is $2k + 1$.

For example, if $n = 16$, then $r = 4$ and $k = 0$, so the expression $2k + 1$ comes out to 1, which means that seat 1 is the winner. If $n = 17$, then $r = 4$ and $k = 1$, so $2k + 1 = 3$, and seat 3 is the winner. If $n = 18$, then $r = 4$ and $k = 2$, so $2k + 1 = 5$, and seat 5 is the winner, and so on, up to $n = 31$, where $r = 4$ and $k = 15$, so $2k + 1 = 31$, and seat 31 is the winner. When we reach $n = 32$, we switch to $r = 5$ and k goes back to 0.

The problem with the rule $2k + 1$ is that it doesn't tell us how to get k in terms of n. Writing k as $n - 2^r$ gets us one step closer, but we still have to describe r. In words, we can describe r as "the biggest whole-number exponent such that 2 to that power is less than or equal to n."

The Greatest Integer Function

There is a standard mathematical notation for the idea just described. The symbol $[x]$ is used to mean "the largest integer that is less than or equal to x." For example, $[3.14]$ is 3, $[5]$ is 5, and $[-6.3]$ is -7. The expression $[x]$ is read as "the greatest integer in x." Many graphing calculators include this as a built-in function. Use your judgment about whether to introduce this notation to your class.

For Your Reference: A Formula

Using the greatest integer notation, we can express r as $[\log_2 n]$, and so $k = n - 2^{[\log_2 n]}$. We can then combine this expression with the formula developed previously to write the formula for the winning seat as

$$2 \cdot \left(n - 2^{[\log_2 n]} \right) + 1$$

Be aware that students rarely develop this general formula on their own, although they may understand its various components.

Explaining the Pattern

Whether or not students come up with a closed formula for the In-Out table, ask them for an explanation of the patterns they have seen and described.

The Power of 2 Case: One way of explaining the rule or pattern is to deal first with the case in which the number of knights is a power of 2. In this case, one cycle around the table eliminates all the knights in the even-numbered chairs. But now the number of knights left is still a power of 2, and the next round still begins with the knight in chair 1.

If we imagine the knights simply "counting off," then once again, the even-numbered knights will be eliminated, the number left will be a power of 2, and we will start with knight 1. If we keep going like this, we will finally end with the smallest power of 2, which is 1. So whenever n (the number of knights) is a power of 2, we end up with the knight in seat 1 winning.

The General Case: The fact that the first knight is the winner if the number of knights is a power of 2 can be used to explain the general case.

For instance, suppose the number of knights is one more than a power of 2; that is, suppose $n = 2^x + 1$. King Arthur begins by calling knight 1 "In" and knight 2 "Out." Now imagine King Arthur being briefly interrupted and then resuming the game. When he returns, there are exactly 2^x knights left, because knight 2 has been removed. When the king resumes, he begins with knight 3, because that's where he left off. In other words, when the king resumes, it's as if he had begun with 2^x knights with knight 3 in the first position. So when $n = 2^x + 1$, knight 3 is the winner.

If the number of knights is two more than a power of 2 (that is, if $n = 2^x + 2$) we can picture King Arthur going "In, Out, In, Out" (for the first four knights) and then pausing. Again, there are exactly 2^x knights left, but in this case, the knight in chair 5 is now the first when the king resumes the game. So when $n = 2^x + 2$, knight 5 is the winner.

In general, if $n = 2^r + k$, the king will eliminate k knights by going through the first $2k$ seats. At that point, there are 2^r knights remaining, and the winner is the knight next in line. This is the knight who was originally in seat $2k + 1$.

Use your judgment as to how much detail to go into here. (A formal proof might use mathematical induction, but that isn't needed here.)

What's It All About?

Intent

This writing assignment gives students a chance to synthesize what they have learned about derivatives.

Mathematics

In this activity, students review the general ideas about derivatives.

Progression

What's It All About? asks students to summarize what they have learned about derivatives. Reading students' work on this activity should give you a good idea of what students have learned about derivatives.

Approximate Time

30 minutes for activity (at home or in class)

15 minutes for discussion

Classroom Organization

Individuals, followed by whole-class discussion

Doing the Activity

Students work on this activity independently.

Discussing and Debriefing the Activity

Have students exchange papers within their groups so that each group member has the opportunity to read at least a couple of other papers, and have each group compile a joint list of ideas about derivatives. Then have groups take turns sharing ideas with the whole class.

We list here some observations that may come up, but this will not be students' final look at derivatives, so don't be concerned if they do not include all of these details. The key idea is that the derivative at a point is a measurement of how the function is changing at that point.

- For a straight line, the derivative is the same as the slope.
- For a curve that is not a straight line, the slope of the "up-close straight line" at a given point is the derivative for that point.
- The "up-close straight line" is called the *tangent line*.

- The derivative tells you the instantaneous rate of change of a function at that point.

- The derivative compares the change in *y* to the change in *x*.

- A negative derivative tells you that the function is decreasing at that point.

- A derivative of zero tells you that the graph is horizontal at that point.

- The derivative of a linear function is the same at every point, while the derivative of a nonlinear function changes from point to point.

- The "larger" the derivative (that is, the larger its absolute value), the faster the function is changing.

Supplemental Activity

Speedy's Speed by Algebra (extension) introduces the formal approach to derivatives, which is much more algebraic than the method presented here. You may want to assign it at this stage in the unit. However, be aware that many students get lost in the algebraic details of this approach and lose sight of the basic intuitive meaning of the derivative. Therefore, use this activity with care.

A Model for Population Growth

Intent

In this section, students work with exponential equations.

Mathematics

These activities establish that an exponential function is the logical choice for modeling population growth. If the relative growth rate for a population were constant, then it makes sense that the absolute growth rate at any point in time would be proportional to the size of the population at that time. This section establishes that a special property of exponential functions is that the derivative at any point is proportional to the value of the function at that point—exactly what is needed to model population growth.

A Model for Population Growth explores exponential equations in a number of contexts, and reviews logarithms through a look at applications of logarithms in science. The development of the derivative continues as well, recognizing the derivative now as a function.

Progression

How Much for Broken Eggs?!!? develops the exponential function in the context of inflation, and then *Small but Plentiful* immediately illustrates that this function appears to be a logical model for population growth as well. *The Forgotten Account* explores the function in yet another context, compound interest.

Two activities refresh students' memories of what they already know about exponential functions and logarithms, beginning with *The Return of Alice*. *The Sound of a Logarithm* reviews logarithms while seeing them at work in the context of the decibel scale.

The Significance of a Sign treats the derivative as a function, setting the stage for introduction of the function notation particular to derivatives.

The main focus of *A Model for Population Growth* is to establish that exponential functions are the best choice for modeling population growth. *Slippery Slopes* develops the principle that the value of the derivative of an exponential function is directly proportional to the value of the function itself. Students see in *How Does It Grow?* that this is not a property of linear or quadratic functions, thus the exponential function is better suited to representing the population growth in the unit problem. Students refine this discovery in *The Power of Powers* and *The Power of Powers, Continued*, as they examine just what form the exponential equation must have to exhibit the proportionality property.

How Much for Broken Eggs?!!?
Small but Plentiful
The Return of Alice
Slippery Slopes
The Forgotten Account
How Does It Grow?
The Significance of a Sign
The Sound of a Logarithm
The Power of Powers
The Power of Powers, Continued

How Much for Broken Eggs?!!?

Intent

This activity introduces exponential functions as a possible model for population growth.

Mathematics

How Much for Broken Eggs?!!? uses inflation to illustrate exponential growth. Students write an exponential function for the situation.

Progression

A brief introductory discussion transitions students into this next section of the unit, regarding exponential functions. The activity then asks students to consider what happens to the price of eggs if the price inflates by 5% each year. Students work in groups and then discuss their results as a class.

Approximate Time

5 minutes for introduction

30 minutes for activity

15 minutes for discussion

Classroom Organization

Small groups, followed by whole-class discussion

Doing the Activity

What's Next—A Brief Transition

Early in the unit, after being introduced to the central problem, students examined linear functions and the concept of slope. They saw that linear functions did not provide a good model for population growth, and they were led to a broader view of rate of change, represented by the derivative. Briefly review these developments in the unit. Tell students that *How Much for Broken Eggs?!!?* will suggest another type of function for them to consider as a model for population growth.

How Much for Broken Eggs?!!?

This activity gives students a chance to see exponential growth in a context other than populations. It requires no further introduction beyond the transition just described.

A Transition from Addition to Multiplication

Most students will probably begin work on this activity by adding 5% to the cost of eggs for each year that goes by. This approach will suffice for Questions 1 and 2 but probably will be inadequate when they attempt Questions 3 and 4. An important element of the activity is getting students to make a transition between this approach and the approach of multiplying by 1.05 for each year, which is arithmetically equivalent. This transition will help them develop a general equation, and the multiplicative approach is more conducive to the idea of an exponential function.

As groups work, you can give them hints to help with this transition or perhaps bring the whole class together for discussion. You might ask, **After one year, the price becomes what percentage of the original price?** Or you can have them express the result of adding 5% in algebraic terms. For instance, ask, **What would the price be after a year if it started at x dollars?** If students write this as $x + 0.05x$, ask, **How can you simplify this?** You can use a phrase like "combine terms" or mention the distributive law to help them come up with the expression $1.05x$. Once students have made this transition to a multiplicative approach, have them return to Question 3.

What about Rounding Off?

Students may ask whether they should round off their answers to the nearest penny each year before computing the next year's price. You can suggest that they experiment with this to see if it matters. It turns out to make very little difference in this case. For example, if you multiply by 1.05 and round off to the nearest penny each year, then the price after ten years is $1.45. If you simply compute $0.89 \cdot 1.05^{10}$, you get $1.4497, to the nearest hundredth of a cent. However, help students recognize that it's easier simply to multiply by 1.05 each year and to round off at the end of the problem if necessary.

Discussing and Debriefing the Activity

Have selected students from several groups report on their groups' answers to Questions 1 and 2. As noted in the introduction to the activity, students might find the answers to these questions by simply adding 5% to the price each year.

Questions 3 and 4

Next turn to Questions 3 and 4. Use the discussion of Question 3 to bring out the general equation $P = 0.89 \cdot 1.05^t$ for the price after t years. Be sure students understand that this equation is based on the fact that adding 5% to the price is equivalent to multiplying it by 1.05. Also, have students explicitly write an equation that could be used to answer Question 4, such as $0.89 \cdot 1.05^t = 100$. (We will refer to this equation in the discussion of *Small but Plentiful*, in the review of logarithms.)

The price of a dozen eggs in the year 2100—that is, after 100 years—is approximately $117.04, and the price will first go over $100 in 97 years, that is, in the year 2097.

The Recursive Approach

Another way to describe the change in the price is to write an equation expressing P_{n+1} in terms of P_n, where P_n is the price after n years. An equation like $P_{n+1} = 1.05P_n$ is called a *recursive formula*. If students develop an equation of this type, you should validate their work and acknowledge that with today's technology, recursive functions are often as useful as closed-form equations. (*Note:* This type of function can be developed using the "add 5%" approach. It would simply state $P_{n+1} = P_n + 0.05P_n$.)

Inflation as an Exponential Function

Ask students, based on the equation $P = 0.89 \cdot 1.05^t$, **What kind of function is this?**, and review the term **exponential** function. Emphasize that this means the variable is being used in the exponent. (*Note:* A more precise definition of the term *exponential function* will be developed later in the unit.)

Ask, **What would the graph of this function look like?** Have students graph the function $y = 0.89 \cdot 1.05^x$ on their calculators to verify their ideas.

Key Questions

After one year, the price becomes what percentage of the original price?

What would the price be after a year if it started at x dollars? How can you simplify this?

What kind of function is this?

What would the graph of this function look like?

Small but Plentiful

Intent

The purpose of this activity is to suggest that if there were no complicating factors, it would make sense to think of population growth as basically exponential.

Mathematics

This activity establishes the exponential function as the logical candidate to model population growth. The discussion of the activity includes a review of logarithms.

Progression

Students work on this activity individually and then discuss their results as a class.

Approximate Time

25 minutes for activity (at home or in class)

30 minutes for discussion

Classroom Organization

Individuals, followed by whole-class discussion

Doing the Activity

This activity uses population growth as an example of exponential growth.

Discussing and Debriefing the Activity

Let selected students share their answers for Questions 1a through 1d. Students should see that the number of creatures keeps doubling.

Then turn to Question 2. If no one can present (and explain) a general formula, return to Question 1d and get a more detailed explanation. Bring out explicitly that 30 days have gone by and that this represents 60 twelve-hour intervals, so the number of creatures has doubled 60 times. Thus, the number of creatures is 2^{60}, which is approximately $1.15 \cdot 10^{18}$ (about 1 quintillion). You can then let groups work to develop a formula. (If a further hint is needed, ask where the number 60 comes from.)

Students will probably write the formula as 2^{2d}, but if someone proposes the equivalent expression 4^d instead, get an explanation for why 2^{2d} and 4^d are equivalent. Point out in either case that the growth of the creature population is described by an exponential function.

Finally, turn to Question 3. Bring out in the discussion that the task here essentially is to find a value of d such that $2^{2d} \geq 1,000,000$. The problem context requires students to find a whole-number solution, and the minimal whole-number solution is $d = 10$. That is, the number of creatures first goes over 1 million ten days after the start of the experiment. Because the experiment began at 12:01 a.m. on January 1, the number of creatures first goes over 1 million at 12:01 a.m. on January 11.

You might connect this discussion to the central unit problem by pointing out that we are talking about population growth once again.

Reviewing Logarithms

Use this opportunity to review the concept of a logarithm. In Question 3, students needed to deal with the inequality $2^{2d} \geq 1,000,000$. Tell them that although this was an inequality and had $2d$ as the exponent, you want to discuss briefly the equation $2^x = 1,000,000$.

Ask students what they can say, based on their work in Question 3, about the exact solution to this equation. If necessary, remind them that they saw that 2^{20} is just over 1,000,000, and help them see that the exact solution to the equation $2^x = 1,000,000$ is a little less than 20.

Then ask students, What is the special name for the solution to the equation $2^x = 1,000,000$? If necessary, suggest using the concept of a logarithm. Remind students that the solution can be written as $\log_2 1,000,000$, and review how to read this expression. (Some people read this as "log, base two, of one million." Others prefer "log of one million to the base two." Either is acceptable.)

To continue the review, ask students, What does $\log_2 8$ mean? You might break this question down into two more specific questions:

- What equation does the expression $\log_2 8$ solve?
- What is the solution to that equation?

Review that $\log_2 8$ is the solution to the equation $2^x = 8$, and identify the solution as $x = 3$. Tie these facts together to establish that $\log_2 8 = 3$.

Then ask students to make up a couple of problems on their own. Focus on the basic meaning of a logarithm. Students might express the meaning of $\log_a b$ in words—for example, "What power should I raise a to in order to get b?"—or in terms of an equation—"What is the solution for x in the equation $a^x = b$?"

Years of Inflation as a Logarithm

Return to Question 4 of *How Much for Broken Eggs?!!?* and review that students wanted to solve the equation $0.89 \cdot 1.05^t = 100$. Ask students how they could express the solution to this equation in terms of logarithms. Students might rewrite the equation as $1.05^t = 100 \div 0.89$, do the division to get that the right side is approximately 112.4, and then express the solution as $t = \log_{1.05} 112.4$.

Note: Students have not seen any way to find the numerical value of this logarithm directly via calculator. It is sufficient that they know what the expression means and realize that the value can be found by solving the equation $1.05^t = 112.4$, which they would most likely do using guess-and-check or a graph.

Key Questions

What is the special name for the solution to the equation $2^x = 1,000,000$?

What does $\log_2 8$ mean?

The Return of Alice

Intent

In this activity, students review principles about exponents and the meaning of logarithms.

Mathematics

This activity brings back the Alice metaphor from the Year 2 unit *All About Alice* in order to review the basic principles for working with exponents as well as the meaning of logarithms.

Progression

The Return of Alice poses a number of relatively straightforward questions in the context of the *All About Alice* metaphor.

Approximate Time

5 minutes for introduction

25 to 30 minutes for activity (at home or in class)

10 to 15 minutes for discussion

Classroom Organization

Individuals, followed by whole-class discussion

Doing the Activity

If you have students who did not use the Year 2 IMP curriculum, you may want to have other students provide a brief summary of the Alice metaphor before assigning this activity.

Discussing and Debriefing the Activity

The amount of time you spend discussing this activity will depend on how well students recall key ideas from the Year 2 unit *All About Alice*.

Question 1 essentially reviews the Alice metaphor, and students should see that the answers are found as exponential expressions using the type of cake as the base and the number of ounces as the exponent.

Question 2 reviews the additive law of exponents. For Question 2a, using the Alice situation to explain the equation $2^4 \cdot 2^3 = 2^7$, students should see that Alice's height is multiplied first by 2^4 and then by 2^3, and that the overall result is to multiply her height by 2^7.

For Question 2b, they should be able to write each expression as an appropriate product of 2's and perhaps develop a display like this:

$$2^4 \times 2^3 = 2^7$$

$$(2 \cdot 2 \cdot 2 \cdot 2) \times (2 \cdot 2 \cdot 2) = 2 \cdot 2 \cdot 2 \cdot 2 \cdot 2 \cdot 2 \cdot 2$$

For Question 3a, students should come up with an Alice situation, such as having Alice eat 6 ounces of base 4 cake each day for a week, explaining that the total amount of cake is 6 · 7 ounces. For Question 3b, they might come up with something similar to the display for Question 2b, perhaps using a shorthand of some type to indicate that there are seven products of the form 4 · 4 · 4 · 4 · 4 · 4 to be multiplied together.

Questions 4 and 5

Use Question 4 to review the meaning of logarithms. That is, students should understand that the expression $\log_2 32$ is *defined* as the solution to the equation $2^x = 32$.

For Question 5, students should get the exponential equations $3^x = 81$, $2^x = 128$, and $5^x = 93$. (Of course, they might use different variables.) Be sure they express the exact solutions as logarithms, as well as give numerical solutions.

Supplemental Activities

Looking at Logarithms (extension) guides students in developing some standard properties of logarithms.

Slippery Slopes

Intent

In this activity, students explore the relationship between the derivative of an exponential function at a point and the y-value at that point.

Mathematics

So far, students have been looking at the derivative as something found at a single point; *Slippery Slopes* requires them to begin to think of the derivative as a function. This activity develops the principle that the derivative of an exponential function at any point is proportional to the y-value at that point. This property is closely related to the suitability of exponential functions as models for population growth, and will also serve as the basis for development of the number e.

Progression

Students work in groups to create tables of values for three simple exponential functions—each table contains values for the exponent, the value of the function at each point, and an approximation of the derivative at that point. By examining the tables, students write an equation for each function that expresses the derivative in terms of the y-value.

Approximate Time

80 minutes for activity

20 minutes for discussion

Classroom Organization

Small groups, followed by whole-class discussion

Doing the Activity

Have students turn to the activity *Slippery Slopes* and read through Question 1. Then have groups complete the table for the value $x = 3$, and go over the result as a whole class. That should make the instructions fairly clear, and you can then let groups continue on their own. If it seems necessary after groups work for a while, bring the class together again to develop a second row for the table.

By the end of the work on this activity, all groups should see that for the function $y = 2^x$, the derivative appears to be some fixed multiple of the y-value. Some groups may discover that this pattern seems to hold for other exponential functions as well.

What's a "Good Approximation"?

In this and subsequent activities, students will encounter many examples in which they need to find the approximate numerical value of a derivative at a particular point. This is a good occasion to discuss how to decide when the approximation is "good" and how to find the numerical value efficiently.

By this time, students should be fairly familiar with the general process. For instance, to find the derivative of the function $y = 2^x$ at the point (3, 8), they probably know that they can compute the y-value that corresponds to a value of x just above 3 and find the ratio of the change in the y-value to the change in the x-value.

They might not, however, be clear on how to decide *how much above* 3 the x-value should be. For instance, they may start with $x = 3.1$, then use 3.01, then 3.001, and so on, computing the appropriate expression of the form $\frac{y_2 - y_1}{x_2 - x_1}$ in each case.
They may compute several such ratios before deciding that their value is "good enough." They may sense correctly that there is no simple rule for knowing when they are "close enough" to the derivative, but you can lead a discussion on an intuitive level about this.

You might suggest that students start right off with a value "very close" to 3 (perhaps 3.001) and then do another that is "even closer" (such as 3.0001). Help them see that if there is very little change in the resulting ratios, then the value they have is probably a "good enough" approximation. For instance, using $x = 3.001$ leads to the ratio 5.547, while using $x = 3.0001$ leads to the ratio 5.545. This should suggest that students can be fairly confident that the exact value is roughly 5.5. (A third computation, using $x = 3.00001$, will confirm that the ratio is essentially the same.)

Point out to students that they can also look at a graph to see that it is "nearly straight" over "short" intervals, so that using much smaller changes in x doesn't have a major effect on the estimate of the derivative. Although you should clarify that this approximation process involves some uncertainty, let them know that this approach will be adequate and lead to fairly accurate results in this unit.

Looking at the Graph

Before analyzing the values in the table, have students look at the graph of the function $y = 2^x$ to verify the reasonableness of the values for the derivatives. Whether they use a hand-drawn graph or a calculator graph, they will need to take into account the scales of their axes. That is, they might choose different scales for the two axes in order to accommodate the fact that the function grows rapidly. As a result, the graph may not appear as steep as the numerical values of the derivative suggest. Even so, they should at least confirm that the steepness is increasing rapidly as x increases.

Hints for Question 1b

While groups work, your main task is to help them focus on the relationship between the derivative and the y-value. Here are some sample values for the function $y = 2^x$, using whole numbers for x:

x-value	y-value	Derivative
0	1	0.693
1	2	1.386
2	4	2.773
3	8	5.545
4	16	11.090

Students may get slightly different values in the third column, because they are getting approximate values of the derivatives, not differentiating symbolically. The values in this table are correct to the nearest thousandth.

If groups do not see the relationship that the derivative is a fixed multiple of y, suggest that they arrange the rows sequentially as shown in our table, and ask, How does the derivative change as x increases? They should see that every time x goes up by 1, the derivative doubles.

One way to help them focus on this is to add a final row to the table, with a "generic" x-value. Ask students, What's the last column if the first column is simply x? If they have found the doubling pattern, they should see that the "initial value" of the derivative, 0.693, has been multiplied x times by 2. Thus, the extended table would look like this:

x-value	y-value	Derivative
0	1	0.693
1	2	1.386
2	4	2.773
3	8	5.545
4	16	11.090
x		$0.693 \cdot 2^x$

Students can then fill in the "generic" *y*-value, 2^x, and see that the derivative is simply 0.693 times the *y*-value. We suggest that you post this table so students can refer to it over the next several days.

Exploring Other Exponential Functions

Once students have found the relationship between the derivative and the *y*-value for the function $y = 2^x$, they should go on to the function $y = 10^x$ and then to other functions of the form $y = b^x$. They will probably see that a similar relationship holds in each case, although the proportionality constant varies.

Students may see that for all such functions, the ratio between the derivative and the *y*-value is equal to the derivative at the point where $y = 1$. (This observation has nothing to do with the specific situation here, but is intrinsic to situations involving direct proportion.) If no groups see this on their own, you can bring it out in the final discussion.

Discussing and Debriefing the Activity

Have students present their results from the activity, using tables as described in Question 1.

Question 1

Even if the class began the discussion of Question 1 previously, review the question thoroughly, having students describe what they discovered. The table for the function $y = 2^x$, using whole numbers for *x*, is repeated here. (See the previous suggestion about adding a row with entries *x*, 2^x, and $0.693 \cdot 2^x$. If this table isn't already posted, you may want to post it now.)

x-value	*y*-value	Derivative
0	1	0.693
1	2	1.386
2	4	2.773
3	8	5.545
4	16	11.090
x	2^x	$0.693 \cdot 2^x$

Although there are various ways to describe the relationships among the columns, focus the class by asking, **What is the relationship between the derivative and the *y*-value?** Students should see that the derivative seems to be approximately 0.693 times the *y*-value. Tell them that although they have not proved this fact, it can be proved that the derivative for this function is actually exactly a fixed number times the *y*-value.

You can point out the meaning of the particular *proportionality constant* 0.693 after the introduction of natural logarithms in the discussion following *The Limit of Their Generosity*.

Questions 2 and 3

Go through a similar discussion for the function $y = 10^x$, for which the corresponding table looks like this:

x-value	*y*-value	Derivative
0	1	2.3
1	10	23
2	100	230
3	1000	2,303
4	10,000	23,026
x	10^x	$2.30 \cdot 10^x$

Here, students should see that the derivative is approximately 2.30 times the *y*-value. You need not fully discuss other functions. But do ask groups to describe their results in a general way. Bring out that in each case, the derivative seems to be proportional to the *y*-value.

The "Proportionality Property"

For convenience, we will say that a function whose derivative is proportional to its *y*-value has the "**proportionality property**." You might introduce this phrase now, and caution students that this is not standard mathematical terminology. (This phrase will be used in *The Power of Powers*.)

Be sure students realize that for a given function, having the proportionality property means that there is a single number *c* such that *for every choice of x*, the derivative at the point with that *x*-value is *c* times the *y*-value that goes with that *x*-value. For the function $y = 2^x$, the value of *c* is about 0.693; for $y = 10^x$, the value of *c* is about 2.30. Ask the class if anyone knows the term for *c*, and review the phrase **proportionality constant**.

You may want to reinforce the idea of the special property by looking at one more example, with an eye to pointing out that the proportionality constant is different for different functions but is fixed, for all *x*, for any given function.

Students may see that the constant c is the value of the derivative at the point where $y = 1$. If this comes up, see if anyone can explain why this is the case. If not, you need not bring it up.

Key Questions

How does the derivative change as x increases?

What's the last column if the first column is simply x?

What is the relationship between the derivative and the y-value?

The Forgotten Account

Intent
In this activity, students examine compound interest.

Mathematics
This activity gives students more experience working with exponential functions, this time in a situation involving compound interest. The discussion again emphasizes a multiplicative approach rather than the "add on the interest" method.

Progression
The Forgotten Account asks students to consider what happens to money left in an account with compounding interest. They will look at compound interest again in *The Generous Banker*.

Approximate Time
25 minutes for activity (at home or in class)

10 minutes for discussion

Classroom Organization
Individuals, followed by whole-class discussion

Doing the Activity
Students work on this activity independently.

Discussing and Debriefing the Activity
Have one or two students explain how they got the answer to Question 1. (The amount in the account after five years is $62.31.) If the presenters use the "calculate the interest and add it on" method for coming up with the year-by-year totals, ask if anyone used a method like that used for the inflation formula in *How Much for Broken Eggs?!!?* If necessary, use hints such as those suggested for that activity to bring out that at the end of each year, the amount in the account is 1.045 times the amount at the beginning of that year.

There is nothing incorrect about the "add on the interest" method, but the multiplicative approach is simpler to use and lends itself much better to developing a general equation.

After the multiplicative approach has been described for Question 1, move on to Question 2. Students' comfort level with the general formula will give you a sense of how well they understand the multiplicative approach.

Questions 3 through 5

Be sure to get the explicit equation asked for in Question 3. Probably, most students will write the equation as $50 \cdot 1.045^t = 500$, although some may divide through by 50 to write $1.045^t = 10$. Either is fine. Students will likely have solved the equation using guess-and-check or a graph. (After 52 years, the account has reached $493.19, and after 53 years, the account has reached $515.39. Because the interest is added only at the end of each year, the exact solution to the equation is not really appropriate to the problem. You might bring out that one might instead ask for the smallest whole-number solution to the *inequality* $50 \cdot 1.045^t \geq 500$.)

Question 5 provides another opportunity to review the meaning of logarithms. Students should be able to express the result as $\log_{1.045} 10$, although this expression does not help them compute the value.

How Does It Grow?

Intent

Students find that ordinarily the growth in a population is proportional to the population itself, so neither a linear function nor a quadratic function makes a good model for population growth.

Mathematics

In this activity, students look at the derivatives of linear and quadratic functions at various points and see how they change. Since it is logical that how fast the population grows should depend on the size of the population, it is evident from the derivatives of these functions that neither of them will serve as appropriate models for population growth. The exponential function, on the other hand, is perfectly suited for modeling population growth, as seen from the proportionality property.

Progression

Students work on the activity individually and then share their results as a class.

Approximate Time

30 minutes for activity (at home or in class)

25 minutes for discussion

Classroom Organization

Individuals, followed by whole-class discussion

Doing the Activity

Students work on this activity independently.

Discussing and Debriefing the Activity

Have students discuss the activity briefly in groups. As you circulate, you can determine how well students understand the mechanics of calculating derivatives. When you bring the class together for discussion, we suggest that you begin with Question 1, then discuss Question 4 along with Question 2, and then turn to Questions 3 and 5.

Question 1

Have a volunteer answer and explain Question 1. Presumably, the student will say that the larger town should grow by 80 people, because it is twice as big as the smaller town. You might get students to expand on this explanation with details such as that there would be twice as many births and twice as many deaths. If it arises, guide students to articulate the general principle that the amount of growth depends primarily on the size of the town.

Questions 2 and 4

For Question 2, have a couple of students each share a specific x-value they chose, the corresponding y-value, and the value of the derivative at that point. You can continue with this until it's clear that the derivative has the same value in every case. Bring out that this essentially answers Question 4a. That is, saying "the derivative is constant" is an appropriate answer to Question 4a.

Ask, **Why is the derivative the same at every point?** Bring out that having a constant derivative (or constant slope) is characteristic of linear functions. Then let students share ideas about Question 4b. Connect the discussion with the example in Question 1. Ask if the rate of growth, according to the function f, is twice as big when the population is 10,000 (that is, for $x = 3332$) as when the population is 5000 (that is, for $x = 1665.33$). They should see that the rate of growth for f (that is, the derivative of f) is constant, so the function f doesn't work the way Question 1 would suggest.

Questions 3 and 5

Next, have several other students each share a combination of results for Question 3, and ask what relationships students found between the derivative and the x-value or y-value. They should see that the derivative appears to be equal to $2x$. (You can assure them that, in fact, the derivative is always $2x$, but be sure they realize that they haven't proved this.)

Ask, **Did you see a relationship between the derivative and the y-value for the function in Question 3?** Students probably will not find any such relationship, and you can simply acknowledge that fact.

Then let students share their thoughts about Question 5b. You might ask, **In population growth, is the rate of increase proportional to the time elapsed?**, in order to demonstrate that a quadratic function like that in Question 3 is not appropriate.

The Nature of Population Growth

Follow up the discussion of the activity by asking, **What 'should' the rate of population growth depend on?** Build on students' conclusions from Question 1. You might also suggest that students look back at the data from the unit problem. Bring out that the population increase per year has been much larger in this century than in previous centuries. You can use a graph of the data to bring out that the slope has increased substantially.

Ask, **Why is the annual increase in population so much larger now than in 1650?** Although students may identify a variety of factors, help them to see that the main reason for this is that there are so many more people to have children. (The number of deaths is also greater, but this still means that the increase, which essentially is calculated as "births minus deaths," is much greater.)

Bring out that the increase in population each year should depend on the size of the actual population. The goal is for students to identify these three ideas:

- The annual population increase should be roughly proportional to the size of the population.
- The population itself is represented by the y-value of the population function.
- The rate at which the population increases is like a derivative.

Together, these ideas mean that the derivative of the population function should be proportional to the y-value. As students found in *Slippery Slopes,* exponential functions have this exact property.

Key Questions

Why is the derivative the same at every point?

Did you see a relationship between the derivative and the y-value for the function in Question 3?

In population growth, is the rate of increase proportional to the time elapsed?

What 'should' the rate of population growth depend on?

Why is the annual increase in population so much larger now than in 1650?

The Significance of a Sign

Intent

In this activity, students explore the significance of the sign of the derivative.

Mathematics

Students' work in this activity will reveal that the derivative of a function indicates the "tilt" of the graph at that point, with a derivative of zero occurring at the turning point.

The discussion following this activity introduces the idea that the derivative of a function is itself a function and has its own function notation. In both *Slippery Slopes* and this activity, students looked at the derivative as a function, rather than as a number to be evaluated at an individual point on a graph. The discussion here brings out the new, broader perspective and introduces two notations commonly used for the derivative function—y' and $f'(x)$.

Progression

In this activity, students indicate the regions of a given graph where the derivative is positive, negative, and zero. They also sketch graphs where the derivative is always positive and where there are exactly two points where the derivative is zero. In the discussion, students see that the derivative is a function, and they learn common notation.

Approximate Time

20 minutes for activity (at home or in class)

20 minutes for discussion

Classroom Organization

Individuals, followed by whole-class discussion

Doing the Activity

Students work on this activity independently.

Discussing and Debriefing the Activity

Let students compare ideas in their groups, and then have a student present Question 1. Students should recognize that the derivative is positive where the function is "going up" (as it goes from left to right) and negative where the function is "going down." They should also see that the derivative is 0 at the "high point" of the graph.

More generally, bring out that the derivative of a function is zero at any point on the graph where the tangent line is horizontal and that this generally occurs at points where the curve switches from "going up" to "going down" (or vice versa). Introduce the term *turning point* for such a point.

Ask, **Can the derivative be 0 without the graph having a turning point?** You may want to use the behavior of the graph of the function $y = x^3$ at (0, 0) to illustrate this, or you can leave it as an open question.

Have two or three students present Question 2. Their graphs should show functions for which the *y*-coordinate increases as the graph goes to the right. (They might use the graph of an exponential function to illustrate this idea. If students give only straight-line examples, press them to come up with other possibilities.)

Finally, have two or three students present Question 3. These will likely be graphs with two turning points (although they need not be).

The Derivative as a Function

Point out to students that in this activity, they considered the derivative for every value of *x*. Similarly, in *Slippery Slopes,* they were concerned about the whole set of derivatives for a given function. For instance, they created a table like this for the function $y = 2^x$:

x-value	*y*-value	Derivative
0	1	0.693
1	2	1.386
2	4	2.773
3	8	5.545
4	16	11.090
x	2^x	$0.693 \cdot 2^x$

Point out that they could have found a derivative to go with any *x*-value, and so they can think of the derivative as depending directly on *x* (essentially ignoring the middle column).

The Notation y′

Tell students that in addition to talking about "the derivative at a point," we also use the word *derivative* to mean this function that associates a rate of change with each *x*-value. Tell them also that this function has a special notation. If the original function is represented using the variables x and y, then the derivative is represented by the notation y', which is read as "y prime." For instance, for the function $y = 2^x$, the table tells us that when x is 3, y' is approximately 5.5.

Ask, **How can you express the special property of this exponential function using this notation?** Help the class develop an equation like $y' \approx 0.693y$, and emphasize that this equation holds true no matter what x is.

The Notation f'(x)

Remind students that the exponential function can be represented in function notation using an equation such as $f(x) = 2^x$. Ask, **How would you represent a point on the graph using function notation?** and, **How might you represent the derivative at this point?** Illustrate the notation $f'(x)$ (which is read as "f prime of x") with an example. For instance, the derivative at the point $(3, f(3))$ is represented by the notation $f'(3)$, so the table shows that $f'(3) \approx 5.5$. In general, $f'(x) \approx 0.693 f(x)$ for the function $f(x) = 2^x$.

Note: In Part II of *Comparing Derivatives,* students will look at the idea that the derivative function, like every function, has a graph, and explore the relationship between the graph of $f(x)$ and the graph of $f'(x)$.

Key Questions

Can the derivative be 0 without the graph having a turning point?

How can you express the special property of this exponential function using this notation?

How would you represent a point on the graph using function notation? How might you represent the derivative at this point?

Supplemental Activities

Finding a Function (extension) extends this activity by having students sketch graphs of functions whose derivatives meet a number of specified conditions.

Deriving Derivatives (extension) guides students to look for an algebraic rule that will give them the derivative of an exponential function.

The Sound of a Logarithm

Intent

In this activity, students work with a logarithmic scale.

Mathematics

This activity gives an example of the use of logarithms in scientific measurement. Sound levels are typically measured with the decibel scale, which uses logarithms to describe the intensity of a sound in comparison to a reference value.

Progression

The Sound of a Logarithm asks students to perform several calculations using the logarithmic equation for decibel level.

Approximate Time

30 to 40 minutes for activity (at home or in class)

10 minutes for discussion

Classroom Organization

Individuals, followed by whole-class discussion

Doing the Activity

Students work on this activity independently.

Discussing and Debriefing the Activity

Let volunteers present answers to the individual questions. Students will probably agree that it's easier to deal with a scale that goes roughly from 0 to 150 than with one that goes from 1 to 1 quadrillion (10^{15}).

Question 1

Question 1 is the most straightforward of the problems, and you can use it to establish the basic method for working with decibels. Students should see that the ratio $\frac{I}{I_0}$ is equal to 100,000 and that the base 10 logarithm of this ratio is 5, so the decibel level is 50.

Question 2

Question 2a reverses the process. Students should see that the ratio $\frac{I}{I_0}$ must have

a base 10 logarithm of 9, so the ratio itself is 10^9, which is 1 billion.

For Question 2b, students might compare the relative intensities for normal conversation and for the whistle (100,000 and 1,000,000,000) to see that the intensity of the whistle is 10,000 times the intensity of normal conversation.

Question 3

This is similar to Question 1, except that the relative intensity is already expressed as a power of 10. Bring out that students can work directly from the exponent to find the decibel level of 130.

Question 4

Questions 4a and 4b are more challenging. On Question 4a, students will probably work through several steps:

- They will find the relative intensity of the first sound, which is $10^{4.2}$, or approximately 15,849.
- They will find the relative intensity of the second sound, which is three times that of the first sound, or approximately 47,547.
- They will find the decibel level of the second sound to be $10 \cdot \log_{10} 47{,}547$, which is approximately 46.8.

Similarly, to answer Question 4b, students will probably find the relative intensities of each of the sounds (which are 10^6 and $10^{6.8}$, respectively, or 1,000,000 and approximately 6,300,000, respectively), and then find the ratio, which is approximately 6.3.

Emphasize that the decibel level of a sound depends on the *ratio* of the intensity of that sound to a certain reference, rather than on an absolute measurement.

Supplemental Activities

Looking at Logarithms (extension) involves general properties of logarithms.

The Power of Powers

Intent

In *The Power of Powers*, students determine which functions with the variable as an exponent have the proportionality property.

Mathematics

In this activity, students explore two functions with exponents that are variables, and determine whether they have the proportionality property that is appropriate to population growth.

Progression

Students work in groups to identify whether given functions have the proportionality property. They will continue this exploration in *The Power of Powers, Continued*.

Approximate Time

5 minutes for introduction

25 minutes for activity

10 minutes for discussion

Classroom Organization

Small groups, followed by whole-class discussion

Doing the Activity

Begin by discussing the introductory definitions in the activity as a whole class, and then let groups work on the specific examples. You might find it productive to assign Question 1 to half the class, and Question 2 to the other half. Alternatively, you could work through Question 1 as a whole class and then have groups work on Question 2.

If groups seem to have trouble getting started, suggest that they pick a specific value of x and get a rough idea of the values of the function and of the derivative. Emphasize that they do not need a very precise estimate of the derivative. (See the subsection "What's a 'Good Approximation'?" in the discussion of *Slippery Slopes*.) For instance, the derivative of the function $h(x) = 2^x + 3^x$ at $x = 1$ could be estimated by the expression

$$\frac{\left(2^{1.01} + 3^{1.01}\right) - \left(2^1 + 3^1\right)}{0.01}$$

To test for the proportionality property, students will have to look at the ratio of the derivative at a point compared to the value of the function at that point, and compare these for various values of x. Have them calculate this ratio at two different points. In Question 1, they should see that the ratios are quite different. They can use a third point to confirm their conclusions.

Discussing and Debriefing the Activity

Have students from each group share their numerical results for Question 1. You might have them put these results in a table like that used for *Slippery Slopes*, like this:

x	$h(x)$	$h'(x)$
1	5	4.7
2	13	12.7
5	275	289

Students should see that the ratio of $h'(x)$ to $h(x)$ is not the same in the three cases (although the ratios are not very far apart). They may want to find ratios for more cases to be confident that the function really does not have the proportionality property.

For Question 2, on the other hand, they might get results like these:

x	$p(x)$	$p'(x)$
1	9.2	6.4
2	18.4	12.8
5	147.2	102.4

In this case, the ratio of $p'(x)$ to $p(x)$ seems to be identical in each case (approximately 0.7), so the function $p(x) = 4.6 \cdot 2^x$ seems to have the proportionality property, with a proportionality constant that is approximately 0.7.

If students notice that the proportionality constant seems to be about the same as for the function defined by the equation $y = 2^x$, you can confirm that, in fact, the proportionality constants are identical.

The Power of Powers, Continued

Intent

This activity continues the work from *The Power of Powers*.

Mathematics

In this activity, students find that all functions of the form $y = k \cdot b^{cx}$ (for appropriate values of b) have the proportionality property.

Progression

Students work on this activity individually, and then discuss their results as a class.

The discussion establishes that only functions of the form $y = k \cdot b^{cx}$ (for appropriate values of b) have the proportionality property.

Approximate Time

30 minutes for activity (at home or in class)

20 minutes for discussion

Classroom Organization

Individuals, followed by whole-class discussion

Doing the Activity

By examining derivatives at various points for each of several more equations, students confirm that all functions of the form $y = k \cdot b^{cx}$ (for appropriate values of b) have the proportionality property.

Discussing and Debriefing the Activity

Question 1

Let different students present their results for Questions 1a through 1c. They should find that the functions in Questions 1b and 1c have the property that the derivative is proportional to the y-value, while the function in Question 1a does not.

Some Sample Tables

Here are sample tables for each of the functions in Question 1, with brief comments on each.

Question 1a: $y = 100 + 2^x$

x	y	y´
0	101	0.693
1	102	1.39
2	104	2.77

Here, the y-values vary only slightly, but the derivatives are changing substantially. For instance, in the first row, y´ is about 0.0069y, but in the second row, y´ is about 0.0136y.

Question 1b: $y = 0.83^x$

x	y	y´
0	1	−0.186
1	0.83	−0.155
2	0.689	−0.128
3	0.572	−0.107
4	0.475	−0.088

If students were thrown off by the fact that the derivatives are negative, use this opportunity to point out that a proportionality constant can be negative. In this example, the function does have the proportionality property, with a ratio of approximately −0.186 between y´ and y.

Question 1c: $y = 2^{3x}$

x	y	y´
0	1	2.08
1	8	16.64
2	64	133.08

Here, the proportionality property holds true, with a proportionality constant of about 2.08.

Question 2

Ask, **What conclusions did you reach about the type of function that has the proportionality property?** and, **Did you make any conjectures that you later discarded?** For instance, they may have initially thought that only functions of the form $y = b^x$, could have the proportionality property. Question 2 from *The Power of*

Powers and Question 1c from *The Power of Powers, Continued* show that this conjecture is false.

Through this discussion, the class should conclude that the proportionality property holds for any function of the form $y = k \cdot b^{cx}$, where c and k are any numbers and b is any positive number. (Students will see later in the unit that this form can be simplified.)

If the class reaches this conclusion on its own, that's great, but if not, you should help students to formulate this generalization. Tell them that the equation

$y = k \cdot b^{cx}$ is the form of the *general exponential function.* (In the discussion of *The Limit of Their Generosity*, we will make this more specific, using the special base e instead of the general base b.) Assure them that all such functions, and *only* such functions, have the proportionality property.

The Proportionality Constant and the Base

Ask, **What does the proportionality constant depend on?** For instance, students may have noticed that the proportionality constant for the function in Question 2 of *The Power of Powers* ($y = 4.6 \cdot 2^x$) was the same as that for the function $f(x) = 2^x$. (You should point this out if students did not notice it.) On the other hand, the function in Question 1c of *The Power of Powers, Continued* ($y = 2^{3x}$) uses the same base but has a different proportionality constant. (Someone might notice that the constant for the function $y = 2^{3x}$ seems to be three times that for the function $y = 2^x$. If this observation comes up, tell the class that it is correct.)

For now, students will need to be satisfied with the statement that the proportionality constant for a function of the form $y = k \cdot b^{cx}$ seems to depend on both b and c.

Key Questions

What conclusions did you reach about the type of function that has the proportionality property?

Did you make any conjectures that you later discarded?

What does the proportionality constant depend on?

The Best Base

Intent

The activities in this section develop the definition of the number *e*.

Mathematics

This phase has two main components, with a central activity for each:

- *A Basis for Disguise:* This activity shows that bases are essentially interchangeable; that is, an exponential function using one number as the base can be expressed using any other number as the base (as long as both numbers are positive and different from 1).

- *Find That Base!*: Once students see that bases are interchangeable, you will point out that using a standard base would make comparisons between functions easier and then ask, "Is there a 'best' base?" That question is resolved by looking for a base in which the ratio of y' to y is actually equal to 1.

Progression

After the activities mentioned above establish that the choice of base for an exponential function is somewhat arbitrary and that a base of approximately 2.7 would make it easier to work with the derivative of the function, students rediscover this handy number in the context of compounding interest. Students first learn how exponential functions can be used as models in the related applications of depreciation (*Blue Book*) and inflation (*Double Trouble*). They then turn to compounded interest in *The Generous Banker* and *The Limit of Their Generosity*.

The latter activity culminates with identification of this special base as the number *e,* and a discussion of natural logarithms and of expressing exponential functions in base *e*.

Solving the central unit problem will require that students be able to fit an exponential equation to the population data. *California and Exponents* helps students extend the curve fitting process they learned earlier to exponential equations, and in *California Population with e's* they revisit this again using base *e*.

A Basis for Disguise

Blue Book

California and Exponents

Find That Base!

Double Trouble

The Generous Banker
Comparing Derivatives
The Limit of Their Generosity
California Population with e's

A Basis for Disguise

Intent

In this activity, students see that exponential functions can be expressed in terms of different bases.

Mathematics

An interesting property of exponential functions is that any exponential function can be expressed in terms of a power of any positive base except 1. This realization, combined with the prior recognition of the proportionality principle, will be the foundation for developing e.

Progression

After a brief review of where we are in terms of the central unit problem, *A Basis for Disguise* leads students to discover through a series of questions that, except for a base of 1, powers of one positive number can all be written as powers of any other positive number.

Approximate Time

5 minutes for introduction

25 minutes for activity

20 minutes for discussion

Classroom Organization

Small groups, followed by whole-class discussion

Doing the Activity

Status of the Unit Problem

Ask, **Where do you stand now in terms of solving the central unit problem?** Guide students to state that they have found that an exponential function would be a natural candidate to model population. They can see this based both on specific situations (such as *Small but Plentiful*) and on general principles (as discussed following *How Does It Grow?*).

Tell students that the next phase of the unit deals with the question, "Which base should we use?"

A Basis for Disguise

Without further introduction, let groups begin the activity *A Basis for Disguise*.

Question 1 will get students started, having them think of 81 as 3^4 and then see 81^5 as $3^4 \cdot 3^4 \cdot 3^4 \cdot 3^4 \cdot 3^4$. Question 2 simply generalizes this idea, and the Alice metaphor should help them visualize what is going on here.

If students need a hint on Question 3, you can review their conclusion in Questions 1 and 2 that 1 ounce of base 81 cake is like 4 ounces of base 3 cake. Then ask what 1 ounce of base 3 cake is worth in terms of base 81 cake.

For Question 4a, be sure students don't get stuck looking for "perfect" answers. For example, in trying to write 7 itself as a power of 5, they should settle for a good approximation, such as $7 \approx 5^{1.21}$, or even $7 \approx 5^{1.2}$. This should lead to a general rule such as $7^x \approx 5^{1.21x}$.

For Question 4b, some students may recognize that there are restrictions on a and b. (They must be positive and different from 1.) However, the main purpose of the question is to allow students to generalize their work from the previous questions in the activity, so focus their attention on the "usual" case, not on the exceptions. (The exceptions will be considered in the discussion of the activity.)

Discussing and Debriefing the Activity

Have one or two students report on their findings for each of Questions 1 through 4a.

Use the presentation and discussion of Question 1 to review the laws of exponents, especially the equation $(A^B)^C = A^{BC}$. Students should see that $81 = 3^4$, and so $81^5 = (3^4)^5$. The key step is recognizing that $(3^4)^5$ is equal to $3^4 \cdot 3^4 \cdot 3^4 \cdot 3^4 \cdot 3^4$ and that this is 3^{20} because there are $5 \cdot 4$ factors of 3 altogether.

Question 2 asks students to generalize this idea, and presenters should come up with the equation $81^x = 3^{4x}$. The reasoning described for the case of 81^5 explains the equation $81^x = 3^{4x}$ whenever x is a positive integer. The Alice metaphor should help students accept that this equation actually holds true for any value of x. (Proving this when x is irrational requires a very formal definition of exponentiation, so you can leave the general principle at an intuitive level.)

Students should see that 1 ounce of base 81 cake is equivalent to 4 ounces of base 3 cake (because $81 = 3^4$), so x ounces of base 81 cake is equivalent to $4x$ ounces of base 3 cake. Using the Alice metaphor, this leads to the equation $81^x = 3^{4x}$.

As suggested in the introduction, students should be able to reason that if 1 ounce of base 81 cake is equivalent to 4 ounces of base 3 cake, then 1 ounce of base 3 cake is equivalent to $\frac{1}{4}$ ounce of base 81 cake. This should lead them to the equation $3^x = 81^{x/4}$ (or equivalently, $3^x = 81^{0.25x}$).

Question 4a

Question 4 is a substantial extension of the ideas of the activity, but the specific example of Question 4a should allow students to see the general principle even if they cannot develop the general rule.

Have the presenters for Question 4a deal with the specific case of writing 7 as a power of 5 before moving on to writing 7^x as a power of 5. For instance, students might write $7 \approx 5^{1.21}$. They should then be able to write the general equation $7^x \approx 5^{1.21x}$.

Ask for the name for the exact number that should be used as the exponent in place of the approximate value 1.21. That is, ask, **What do you call the solution to the equation $5^t = 7$?** As needed, remind students that this type of equation essentially *defines* logarithms, so that the desired value of t is simply $\log_5 7$.

Question 4b

Let volunteers share their ideas on Question 4b. The key idea is that if b itself can be written as a power of a, then there is a general rule for writing b^x as a power of a. More specifically, if $b = a^r$, then $b^x = a^{rx}$. (Clarify that a and b here are both positive numbers.) Although some students may be able to develop the general formula $b^x = a^{(\log_a b)x}$, you need not push for this.

Once students understand the principle behind the "usual" case, ask if there are any positive values for a and b for which this change-the-base method would not apply. Help students see that the cases in which a or b is 1 are exceptions because all powers of 1 are equal to 1.

Summary: Interchangeability of Bases

Before students begin *Find That Base!*, it's important that they see the "big picture" that should emerge from their work in *A Basis for Disguise*. Ask if anyone can summarize the conclusions from this activity, and try to elicit a statement like this:

> **For any positive bases a and b, different from 1, any power of a can be rewritten as a power of b, and vice versa.**

If you want to bring out another approach to this principle, you can use the graphs of exponential functions to do so. You can have the class pick any positive number b other than 1 as the base, and simply ask what values the expression b^x can achieve. (You might review the phrase *range of a function* in this context.) Bring out that no matter what b is, the range is the set of all positive numbers.

Also, remind students that exponential functions seem to present a good model for population growth and that this is why they are looking at them in such detail. Review that the general exponential function was expressed previously using the equation $y = k \cdot b^{cx}$.

Ask, **What does *A Basis for Disguise* say about this general form?** Bring out that a given function can actually be written using any positive base (except 1) simply by appropriately adjusting c, the coefficient of x in the exponent. Students will pursue this idea further in upcoming activities.

Key Questions

Where do you stand now in terms of solving the central unit problem?

What do you call the solution to the equation $5^t = 7$?

What does *A Basis for Disguise* say about this general form?

Blue Book

Intent

In this activity, students recognize that depreciation can be modeled by an exponential function.

Mathematics

For the first time in this unit, students explore an exponential function that is decreasing rather than increasing.

Progression

Students work individually to find the value of a depreciating car after various numbers of years, and to write a formula for the value of the car after t years. The discussion brings out that if two quantities depreciate at the same relative rate, then the ratio of their values stays the same.

Approximate Time

25 to 30 minutes for activity (at home or in class)

15 to 20 minutes for discussion

Classroom Organization

Individuals, followed by whole-class discussion

Doing the Activity

The activity asks students to find the value of a car after various numbers of years using the rule that it depreciates by about 30% each year, and to write a formula for the value of the car after t years.

The second question notes that this means that the value loss of a more expensive car expressed in dollars will be greater than that of a less expensive car and asks whether this means it will eventually be worth less than the cheaper car.

If students ask about the title of this activity, explain that there is an official index to the value of used cars, called the "blue book." When people are considering buying or selling a used car, they will often use the car's "blue-book value" as the basis for setting the price.

Discussing and Debriefing the Activity

Question 1

We suggest that you let a volunteer present a solution for this problem. The biggest difficulty with the problem will probably be in seeing that "losing 30% of the value" is the same as multiplying the value by 0.7 (which is 1 – 0.3).

You can use the specific examples in Question 1a to bring this out. You may also find it helpful to refer students to their work from *How Much for Broken Eggs?!!?*, because in that problem they described 5% inflation as multiplying prices by 1.05 (which is 1 + 0.5). Thus, they should see that the price of the car after t years would be $15,000 \cdot 0.7^t$.

Question 2

Students should realize that Tara is mistaken. They should be able to explain that because each car's value is multiplied by 0.7 each year, the more expensive car is always worth twice as much as the cheaper car. That is, because the two values depreciate at the same percentage rate, the ratio of their values stays the same. Looking at the values of the two cars after one or two years should help clarify this.

Relative Versus Absolute Growth

This is a good context in which to discuss the distinction between *absolute* and *relative* growth. (Relative growth is sometimes called "percentage growth.")

Bring out that the two cars lost different amounts of value but the same percentage of their value. Ask, **What terms are used to distinguish between 'amount of change' and 'percentage of change'?** Review the terms absolute growth rate and relative growth rate as needed. (You may need to clarify that in Blue Book, these are rates of decrease rather than rates of increase.)

To help students distinguish between absolute and relative growth, you might point out that an absolute growth rate is usually expressed as a simple number (per year or other unit of time), while relative growth rate is usually expressed as a percentage (per unit of time).

An example using population growth might help. For instance, if a population grows from 10,000 to 11,000 people in a year, the absolute growth rate is 1000 people per year. The relative growth rate is 10% per year, because 1000 people is 10% of 10,000.

Ask, **If the absolute growth rate is constant, what kind of function do you get?** As a hint, you can paraphrase by asking, **In other words, what kind of function is it if the same amount is added to the price each year?**

Then ask, **If the relative growth rate is constant, what kind of function do you get? In other words, what kind of function is it if the price is multiplied by the same amount each year?** As a further hint, ask students to look at their result for Question 1, or go back to the situation in *How Much for Broken Eggs?!!?* They should see that constant relative growth leads to an exponential function.

To summarize, bring out that we have two different types of constant growth:

- Linear functions are used to describe constant *absolute* growth.
- Exponential functions are used to describe constant *relative* growth.

More on Population

After establishing the distinction between absolute and relative growth, you might want to have the class reexamine the key principle for population growth. Specifically, the growth rate—that is, the *absolute growth rate*—should be proportional to the population. Students should see that this is the same as saying that the relative growth rate is constant.

To look at this numerically, you can have the class imagine a country whose 2000 population is twice its 1950 population. Ask, **Would you expect the same number of births in both years?** Students should see that they would expect about twice as many births when the population was twice as big. (There would also be twice as many deaths, but the absolute growth, which is the difference between births and deaths, would still be twice as big in 2000 as in 1950.)

If another example is needed, you can have students look at the situation involving microscopic creatures (*Small but Plentiful*). There, too, as the population doubled, the number of new creatures in a time interval of a given length also doubled.

Key Questions

What terms are used to distinguish between 'amount of change' and 'percentage of change'?

If the absolute growth rate is constant, what kind of function do you get? In other words, what kind of function is it if the same amount is added to the price each year?

If the relative growth rate is constant, what kind of function do you get? In other words, what kind of function is it if the price is multiplied by the same amount each year?

Would you expect the same number of births in both years?

California and Exponents

Intent

In this activity, students find an exponential function through two given points.

Mathematics

This activity introduces students to the task of fitting an exponential function to data. They will return to this situation in *California Population with e's*.

Progression

An introductory discussion reminds students of their approach in the unit *Meadows or Malls?* to fitting a linear equation to given points. In this activity, they fit an exponential function to two points, then use this equation to answer questions.

Students need to save their results from this activity for use in a later activity.

Approximate Time

10 minutes for introduction

25 minutes for activity (at home or in class)

10 minutes for discussion

Classroom Organization

Individuals, followed by whole-class discussion

Doing the Activity

California and Exponents gives students points that represent populations at two times during the California gold rush, and asks students to find numbers a and b such that the exponential function $y = a \cdot b^x$ goes through both points. A hint suggests they do this by first substituting the coordinates of the first point into the equation. The second question asks them to use their function to predict the population in 2000, and the third question asks whether the population growth has slowed or speeded up since the Gold Rush.

In the unit *Meadows or Malls?*, students learned a method for finding coefficients in a linear or quadratic function by substituting the coordinates of points that were known to fit the function. For instance, in the discussion of *Fitting a Line,* they were looking for equations of the form $y = ax + b$ whose graphs went through the point (1, 2). They saw that if the graph went through (1, 2), then substituting 1 for x and

2 for y had to fit the equation. That is, the coefficients a and b had to satisfy the condition $2 = a \cdot 1 + b$.

Review this idea, perhaps using the example just mentioned. We recommend that you limit this introductory discussion to the case of functions of the form $y = ax + b$. This will allow students to make the necessary generalization on their own to apply this idea to an exponential function.

Discussing and Debriefing the Activity

Let a volunteer present Question 1. The hint shows that a is simply 92,600, and you can use the hint as an opportunity to review the meaning of 0 as an exponent. If students continue using the hint, they should come up with an equation such as $380,000 = 92,600 \cdot b^{10}$, which they can simplify to get something like $b^{10} = 4.10$.

(The number 4.10 is the approximate value of the fraction $\frac{380,000}{92,600}$.)

Students may use a variety of methods to solve this equation, such as guess-and-check or raising 4.10 to the power 0.10. (The value of b is approximately 1.15.)

Point out that there is a unique solution for a and b, just as there is a unique *linear* function that goes through the two given points. Without getting overly specific, try to bring out the parallel between the roles of a and b in the form $y = a \cdot b^x$ and the roles of the same variables in the form $y = a + bx$ for a linear function. In each case, a represents the starting value and b defines the rate of growth. The only distinction is that for linear functions, b is the *absolute* rate of growth and we add b for each unit change in x, while for exponential functions, b is the *relative* rate of growth and we multiply by b for each unit change in x.

Questions 2 and 3

For Question 2, students should see that they need to substitute 150 for x. That is, using the estimate given earlier for b, they need to evaluate the expression $92,600 \cdot 1.15^{150}$. This expression comes out to approximately $1.2 \cdot 10^{14}$, (although a more accurate value of b yields an answer of approximately $1.5 \cdot 10^{14}$). In either case, students should see that if the same relative growth rate had continued, the population of California in 2000 would have been roughly 120 trillion people, so the relative growth rate apparently has slowed down considerably since the Gold Rush period.

Tell students that they will return to the data from this problem in a later activity and that they should save the values they found for a and b.

Find That Base!

Intent

In this activity, students begin the development of *e*.

Mathematics

Students saw in *A Basis for Disguise* that any power of a particular positive base can be rewritten as a power of any other positive base, with the exception of a base of 1. They also saw in *Slippery Slopes* that exponential functions have a proportionality property such that at any point the derivative of the function is proportional to the value of the function itself.

This activity suggests that, in the interest of standardization, it would be nice if all exponential functions could be expressed in terms of a base that was chosen to make the proportionality constant 1. In this way the derivative of the function would be equal to the function itself at every point. That base is known as *e*, although students do not learn that until the discussion following *The Limit of Their Generosity*.

Progression

Find That Base! asks students to estimate the value of *b* for which the derivative of the function is equal to b^x at every point on the graph of the function. It is not necessary that students find a particularly accurate estimate, and students are not yet told that the number they are finding is known as *e*.

Approximate Time

10 minutes for introduction

25 minutes for activity

5 minutes for discussion

Classroom Organization

Small groups, followed by whole-class discussion

Doing the Activity

Review with the class the basic conclusion from *A Basis for Disguise:*

> **For any positive bases *a* and *b*, different from 1, any power of *a* can be rewritten as a power of *b*, and vice versa.**

This means that any exponential function can be expressed using any positive base we like (other than 1). Point out that changing bases involves adjusting the coefficient of *x* in the exponent.

Tell students that scientists like to standardize ways of writing things, because that makes comparisons and estimates easier. Therefore, they like to write all exponential functions with the same base. *A Basis for Disguise* showed us that it's possible to rewrite exponential functions with any given base.

The Proportionality Property and Choosing a Base

You may want to remind students that we are taking time to explore different bases because exponential functions are natural candidates for modeling population growth.

Ask, **What special property makes exponential functions appropriate for modeling population growth?** Review the proportionality property investigated in *Slippery Slopes*. In that activity, students saw that the function $y = 2^x$ had a derivative at each point that was about 0.693 times its *y*-value at that point, and that the function $y = 10^x$ had a derivative at each point that was about 2.3 times its *y*-value at that point.

Ask, **What proportionality constant do you think would be best?** If they don't guess, tell them that scientists prefer a proportionality constant of 1. In other words, they prefer the base for which the derivative at each point is actually equal to the *y*-value at that point.

You can tell the class that scientific work often involves derivatives and that having this constant equal to 1 simplifies the algebra of working with derivatives considerably. Thus, for convenience and ease of comparison, scientists generally express all exponential functions in terms of that special base.

Discussing and Debriefing the Activity

Let students report on their findings. As noted, they need not have found the precise base, but they should understand what they are looking for and recognize that they can find as precise a value as they need to. In fact, for now, it's fine if they simply see that the number they are searching for is approximately 2.7. (*Note:* The use of the symbol *e* for this base will be introduced following the discussion of *The Limit of Their Generosity*.)

Tell students that they will come back to this special base in a few days, but that first they will do something that probably seems totally unrelated. (A wonderful thing about mathematics is how two things that seem totally different can turn out to be related.)

Key Questions

What special property makes exponential functions appropriate for modeling population growth?

What proportionality constant do you think would be best?

Double Trouble

Intent

This activity is intended to build students' intuition about compounding.

Mathematics

In this activity, students find the doubling time for different rates of growth, and find a pattern often known as "the rule of 72": for rates of growth in the 5% to 10% range, $\frac{72}{x}$ approximates the number of years it will take to double at a rate of x%.

Progression

Students work individually to explore doubling time for various rates of growth, and look for a pattern.

Approximate Time

30 minutes for activity (at home or in class)

15 minutes for discussion

Classroom Organization

Individuals, followed by whole-class discussion

Doing the Activity

Double Trouble asks students to find the doubling time with an inflation rate of 5% per year and then to do it again with several other inflation rates, tabulating the results and looking for a pattern or rule. The final question asks how the quadrupling time compares with the doubling time.

Discussing and Debriefing the Activity

Have a volunteer describe how he or she got the doubling time for the 5% inflation rate. Students should see they need to solve the equation $1.05^t = 2$ (to the nearest hundredth, according to the instructions), which is a straightforward calculator task using guess-and-check. If this seems clear to the class, you can then simply have other students give their results and enter them in an In-Out table.

Here are some rows for the table. (For convenience, we've included both doubling and quadrupling time here.)

Inflation rate	Doubling time (in years)	Quadrupling time (in years)
1%	69.66	139.32
2%	35.00	70.01
4%	17.67	35.35
5%	14.21	28.41
10%	7.27	14.55
20%	3.80	7.60

Patterns and Explanations for the Doubling Time

Once you have half a dozen or so rows of the doubling-time table, ask, **What patterns or rules did you find for this doubling-time table?** Students should certainly see that as the inflation rate goes up, the doubling time goes down. (Don't neglect this obvious relationship.)

If students do not offer any other ideas, ask, **How does the doubling time for 5% compare to that for 10%?** Students should see that it's about half, and you can reinforce this relationship by looking at other examples (such as 1% and 2%).

Ask, **Why is the doubling time for 5% about double that for 10%? Is it exactly double?** Students should see that two years of 5% inflation will raise prices a bit over 10%, so the doubling time for 5% is slightly less than double that for 10%.

Other Ideas

Some students may see that this "almost inverse" relationship can be expressed approximately by a formula. For instance, they may see that if the inflation rate is x percent, the doubling time (in years) is approximately equal to $\frac{70}{x}$. (This is not a precise rule, and as x increases, the numerator needs to go up to give more accurate values. This idea is commonly referred to as "the rule of 72" because the expression $\frac{72}{x}$ gives good approximations when the inflation rate is between 5% and 10%.) If students suggest a rule like this, be sure they see that no simple inverse expression gives the exact values.

Some students may express the In-Out rule in terms of logarithms. For example, with a rate of 5%, finding the doubling time means finding the value of t for which $(1.05)^t = 2$. The solution to this equation is, by definition, $\log_{1.05} 2$. In general, the doubling time for an inflation rate of x percent is $\log_{(1 + .01x)} 2$ [and the quadrupling time is $\log_{(1 + .01x)} 4$].

Question 5: Doubling and Quadrupling

Next, turn to Question 5. Ask, **What are the quadrupling times? How do they compare to the doubling times?** Students should notice that the time needed for prices to quadruple seems to be twice the time needed for prices to double. (The rounding may obscure this.)

Ask, **Are the quadrupling times exactly twice the doubling times or is this simply an approximation? Why?** Although rounding may make this relationship appear to be merely an approximation, help students to see that the relationship is exact. They might explain the relationship by saying that if prices double in x years, they will double again in another x years, so they will quadruple in $2x$ years.

Key Questions

What patterns or rules did you find for this doubling-time table?

How does the doubling time for 5% compare to that for 10%?

Why is the doubling time for 5% about double that for 10%? Is it exactly double?

What are the quadrupling times? How do they compare to the doubling times?

Are the quadrupling times exactly twice the doubling times or is this simply an approximation? Why?

The Generous Banker

Intent

In this activity, students continue to examine relative growth, this time in the context of compound interest.

Mathematics

In this activity, students investigate the results of a bank account that compounds interest every twenty years, 1 year, 6 months, and 3 months. This will lead to development of *e* in an upcoming activity.

Progression

Students work in groups on the activity. The subsequent discussion introduces the vocabulary of compound interest, such as *compounded quarterly.*

As this investigation of compounded interest continues in *The Limit of Their Generosity*, it will lead to development of the number *e*.

Approximate Time

25 minutes for activity

20 to 25 minutes for discussion

Classroom Organization

Small groups, followed by whole-class discussion

Doing the Activity

The Generous Banker continues the exploration of relative growth. You may want to assure the class that this idea will soon connect back to the unit problem, but no other introduction is needed.

Discussing and Debriefing the Activity

Have different students give and explain the specific results on each of Questions 1 through 3. Here are the answers, to the nearest penny:

- Question 1: $2,653.30
- Question 2: $2,685.06
- Question 3: $2,701.48

Presenters will probably use the expressions 1.05^{20}, 1.025^{40}, and 1.0125^{80} to find the answers to these questions. In preparation for *The Limit of Their Generosity*, it will be helpful to have students clarify where these bases are coming from. For instance, on Question 2, help students to see that the base 1.025 can be thought of as $1 + \frac{1}{40}$ and that the denominator 40 comes from dividing the 100% interest into 40 separate payments.

Ask, **Why did Adam end up with more than $2,000?** Students should be able to tie this problem in with the activity *How Much for Broken Eggs?!!?* Make sure they realize that something like the practice described in this problem is actually done by banks. Review the term *compound interest* and introduce phrases such as *compounded quarterly.* (The term compound interest was introduced in *The Forgotten Account.*)

Work out a simple problem or two in which students figure out the consequences of compounding. For example, if 10% annual interest is compounded quarterly, then the interest for each quarter is 2.5%. Thus, the principal is multiplied by 1.025 each quarter, and after a year, it is multiplied by $(1.025)^4$, which is approximately 1.1038. In other words, the customer gets approximately 10.38%, rather than 10%, because of compounding.

Key Question

Why did Adam end up with more than $2,000?

Supplemental Activity

The Reality of Compounding (reinforcement) provides a good follow-up to this discussion. It asks students what interest, compounded annually, would actually double the customer's money in 20 years.

Comparing Derivatives

Intent

This activity continues students' work on the meaning of derivatives.

Mathematics

In this activity, students relate derivatives to the graphs of functions and find the graph of the derivative of a function from the graph of the function.

Progression

The first part of *Comparing Derivatives* gives students three functions whose graphs all pass through the same pair of points. It asks the students to graph all three on the same set of axes. Finally, students tell which function has the largest derivative at each of the two points based on the graph and then find the derivatives at those points to confirm their conjecture. The discussion of this part emphasizes that "larger derivative" means "steeper graph."

Part II presents the graph of an unknown function and asks students to sketch the graph of the derivative of the function. The discussion of Part II focuses on the connection between zeroes of the derivative and turning points of the function.

Approximate Time

5 to 10 minutes for introduction

30 minutes for activity (at home or in class)

15 to 25 minutes for discussion

Classroom Organization

Individuals, followed by whole-class discussion

Materials

Optional: Copies of the *Comparing Derivatives* blackline master of the Part II graph (1 per student)

Optional: Transparencies of the *Comparing Derivatives* blackline masters

Doing the Activity

You may want to give one or two students blank overhead transparencies ahead of time so that they can prepare presentations on the graphs in Question 1.

Question 5 of this activity is quite challenging. You may want to suggest to students that they focus on the sign and general magnitude of the derivative. For instance, their graphs of *f'* might illustrate that the derivative is "very positive" at one point or "slightly negative" at another.

Discussing and Debriefing the Activity

Graphs of the functions in Part I and Part II are provided in the *Comparing Derivatives* blackline masters. You may want to make overhead transparencies of these graphs for use in this discussion.

Part I: Shared Points

If you had students prepare graphs for Question 1, let them present their work and discuss Question 2 as well. Otherwise, have a volunteer present the graphs. The three graphs are shown together here. (A copy of this graph is in the *Comparing Derivatives* blackline masters.)

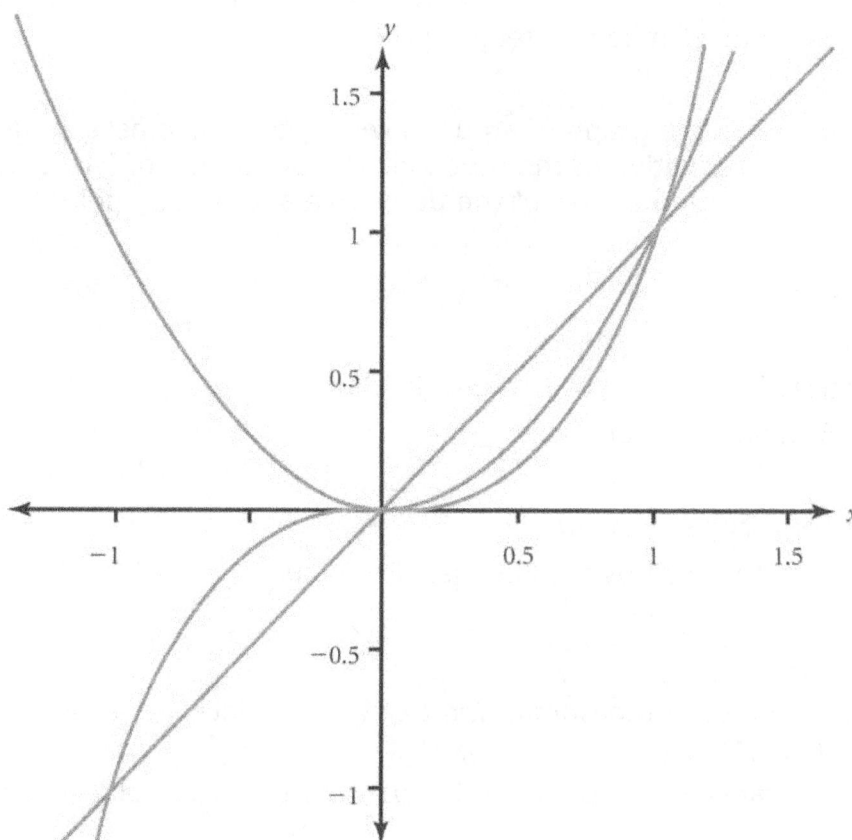

The main goal of Question 2 is for students to review the idea that "larger derivative" means "steeper graph." Reasonably accurate graphs should show that at (1, 1), the graph of $h(x) = x^3$ is the steepest of the three, so it has the largest derivative, while at (0, 0), $f(x) = x$ is steepest, so it has the largest derivative.

The derivatives of $g(x) = x^2$ and $h(x) = x^3$ at the point (0, 0) provide an opportunity to review what it means for a derivative to be equal to 0. In the case of $g(x) = x^2$, (0, 0) is a turning point of the graph. (You can review this term, introduced in the discussion of *The Significance of a Sign*.) In the case of $h(x) = x^3$, there is something more subtle going on—the graph is "flattening out," but then it picks up again.

As needed, discuss the numerical computation of derivatives in Question 3 to be sure students are comfortable with this.

Part II: Derivative Sketch

Begin with a discussion of where the derivative is positive, negative, and zero. This is best done visually, using an overhead transparency of the graph (provided in the *Comparing Derivatives* blackline master).

Then turn to Question 5. Results are likely to vary somewhat because students will need to estimate the derivatives, but the sketches of the derivative function should have the same general shape. Roughly, the derivative graph should look like this, with the derivative being zero at approximately −2.2, 0.6, and 3.8. The main idea is for students to be able to look at the graph of $f(x)$ and determine the sign of $f'(x)$.

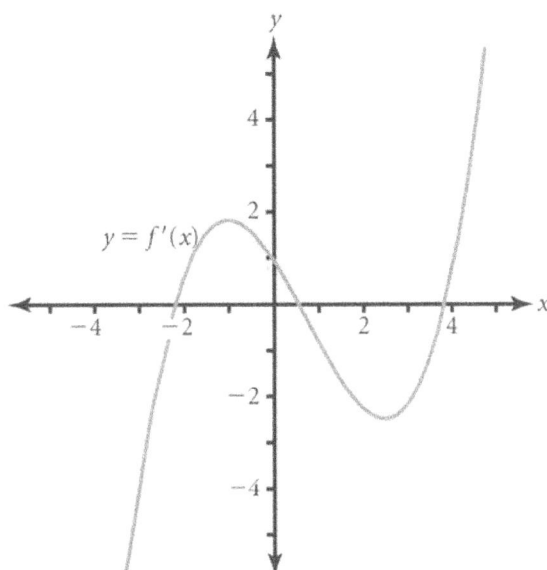

Optional: The Derivative and Concavity

If you want to pursue this further, you can ask why the derivative graph has turning points at about $x = -1.0$ and $x = 2.5$. Students might say something about the graph being "steepest" at these points, so the derivative is at its greatest there (in absolute value). Others might connect these x-values with a change in concavity.

The Limit of Their Generosity

Intent

This activity culminates with defining *e.*

Mathematics

The Limit of Their Generosity is a lead-in to the discussion of the number *e.* The number arises here in connection with compound interest and is defined as the limiting value of the expression $\left(1 + \dfrac{1}{n}\right)^n$, but the discussion will connect it to the "special base" students sought in *Find That Base!*

The discussion following this activity clarifies that if $f(x) = e^x$ then $f'(x) = e^x$ also. It also defines natural and common logarithms. The discussion concludes with a consideration of how to convert an exponential function to base *e.*

Progression

Students work on this activity individually and share their results as a class. The subsequent discussion defines *e,* and students learn how to use the function $y = e^x$ on their calculators. They also learn about natural and common logarithms, and convert exponential functions to base *e.*

Approximate Time

20 to 25 minutes for activity (at home or in class)

50 minutes for discussion

Classroom Organization

Individuals, followed by whole-class discussion

Doing the Activity

The activity continues with the compounding interest situation from *The Generous Banker.* Students evaluate further plans, now calculating the effects of daily and hourly compounding for the twenty-year period.

Discussing and Debriefing the Activity

The computations in this activity should be fairly routine after *The Generous Banker,* but it is important for students to see the results. You can ask for volunteers to share their guesses on Question 1, and then have selected students present results on Questions 2 and 3.

For Question 2, the work should go something like this: There are 365.25 · 20 days in 20 years, which is a total of 7305 days. Therefore, Adam's money is multiplied by $\left(1+\dfrac{1}{7305}\right)^{7305}$ which is approximately 2.71810, and he would end up with about $2,718.10. For Question 3, the interest is given in 175,320 installments (because 7305 · 24 = 175,320), so the multiplying factor is $\left(1+\dfrac{1}{175,320}\right)^{175,320}$. This factor is approximately 2.71827, so Adam gets an additional 17¢ by getting interest every hour.

You might ask the class to go one step further, and compute the result for compounding every minute. It turns out that this gives almost the same result as Question 3, differing by about a penny. (*Note:* Students may run into rounding errors on their calculators when they do some of these computations.)

Students will probably be surprised that the results don't really change much after the first stage or two, even as the frequency of compounding increases enormously. You may want to combine their results from *The Generous Banker* and this activity into a table like this so that they see the numbers more dramatically:

Compounding interval	Number of intervals in 20 years	Amount of interest each interval	Final amount after 20 years
20 years	1	100%	$2,000.00
1 year	20	5%	$2,653.30
6 months	40	2.5%	$2,685.06
3 months	80	1.25%	$2,701.48
1 day	7305	≈0.0137%	$2,718.10
1 hour	175,320	≈0.00057%	$2,718.27
1 minute	10,519,200	≈0.0000095%	$2,718.28

You can tell students (though they may not quite believe it) that no matter how small the intervals get, there is a maximum that can be achieved. To the nearest penny, this maximum is $2,718.28.

By now, students should be reasonably comfortable with expressions of the form $\left(1+\dfrac{1}{x}\right)^{x}$. You can give them a few minutes to investigate this expression using very large numbers for x.

Optional: You can tell students that mathematicians describe the process of seeing what happens as x gets larger and larger as the phrase "taking the limit as x approaches infinity," and perhaps show them the notation $\lim_{x \to \infty} \left(1 + \dfrac{1}{x}\right)^x$.

Defining e

Tell students that the special number that is the upper limit of this compounding process is called *e*. (This is one of several ways to define *e*.) To five decimal places, *e* is equal to 2.71828.

Historical note: The number *e* was given this name by the famous Swiss mathematician Leonhard Euler (1707–1783), who was also responsible for promoting the use of the symbol π as the ratio of the circumference of a circle to its diameter.

Using e on a Calculator

Tell students that the number *e* is so special that the function $y = e^x$ has its own place on most scientific and graphing calculators. Give students a few minutes to see how to work with this function on their calculators. For instance, have them get *e* itself as the result of the expression e^1.

Then remind students of their work on *Find That Base!*, and ask, What do you suppose the special base is from *Find That Base!*? They will probably take the hint and guess that e is that special base.

Ta-da!—The Derivative of y = e^x

As the culmination of students' work since *Slippery Slopes*, have them now work in groups to find the derivative for the function $y = e^x$. (You can assign specific values of x to different groups.)

They should see that for this function, the derivative is actually equal to the *y*-value (at least, to within the level of approximation they can achieve). Emphasize that *e* is the *only* base for which this is true and that this is the main reason the number *e* is so important.

Exponential Functions in Terms of e

Tell students that because of *e*'s special property, functions like $y = 2^x$ and $y = 10^x$ are often written in the form $y = e^{cx}$, for some appropriate choice of *c*. Have groups try to write the function $y = 2^x$ in this form.

If a hint is needed, ask, **How can you write 2 itself as a power of e?** They will probably solve the equation $e^r = 2$ by guess-and-check and come up with an approximate solution of 0.693 for r.

Help students go from the specific equation $2 = e^{0.693}$ (if they start with that) to the general relationship $2^x = e^{0.693x}$. (The Alice metaphor may be helpful. If 1 ounce of base 2 cake has the same effect as 0.693 ounces of base e cake, then x ounces of base 2 cake have the same effect as $0.693x$ ounces of base e cake.) Bring out that an equation similar to $2^x = e^{0.693x}$ can be developed for writing b^x in the form e^{cx} for any positive base b other than 1. (This is the essential idea in *A Basis for Disguise.*)

The General Exponential Function, Revisited

Review with students that earlier in the unit, they used the form $y = k \cdot b^{cx}$ to represent the general exponential function (see *The Power of Powers, Continued*). They saw that functions of this form have the proportionality property; that is, their derivatives are proportional to their y-values.

Ask, **How can this general form be simplified using e?** You may need to guide students to an understanding of how two parameters can be combined into one. They already know that they can write the base b as e to some power (say, $b = e^r$), so that $k \cdot b^{cx} = k \cdot (e^r)^{cx}$. Help them to make the transition from this to $k \cdot e^{rcx}$ and then to see that the coefficient rc is simply a "generic" number, so the most general exponential function is of the form $y = k \cdot e^{cx}$. (You may wish to point out here that x and y are variables; e is a particular number; and c and k are constants, to be determined.)

The Natural Logarithm

Go back to the specific example $2^x = e^{0.693x}$ and ask, **Does anyone recognize the number 0.693 from earlier in the unit?** Bring out that this was the proportionality constant between y' and y for the function $y = 2^x$ (in *Slippery Slopes*). Tell the class that this is not an accident. That is, the coefficient of x in the equation $2^x = e^{0.693x}$ turns out to be the exact proportionality constant between y' and y for the function $y = 2^x$.

Emphasize that in *Slippery Slopes*, the coefficient 0.693 was an approximation. Ask, **How can you find the exact proportionality constant? What do you call the exact solution to the equation $e^c = 2$?** Emphasize that they are looking for a number c for which 2^x is exactly equal to e^{cx}. Help the class to see that this number c is the solution to the equation $e^c = 2$ (this is the case $x = 1$) and that this equation defines $\log_e 2$.

Tell students that because e is often used as a base, scientists also often use logarithms to the base e. (As students have just seen, such logarithms are used to convert exponential expressions to base e.)

Tell the class that these logarithms have a special name and notation. Logarithms using the base e are called **natural logarithms**, and $\log_e a$ is written $\ln a$. Tell the class that "ln" is read as "ell en." That is, we simply name the two letters in the abbreviation. (It may help if you explain that the "l" stands for logarithm and the "n" for natural, and that perhaps they are in reverse order because, in French, one would say "logarithme naturel.")

Finally, tell students that most scientific and graphing calculators have a special key devoted to natural logarithms. Then have them figure out how to calculate $\ln 2$ on their calculators.

Base 10 Logs Are Special, Too

Remind the class that base 10 logarithms are also special, because our number system uses powers of 10. Illustrate that the base 10 logarithm of a number can be useful in estimating the number (and vice versa). For example, if $\log_{10} x = 3.47$, then $x = 10^{3.47}$, so x is between 10^3 and 10^4, that is, between 1000 and 10,000. Similarly, a number between 100,000 and 1,000,000 has a base 10 logarithm between 5 and 6.

Because base 10 logarithms give an easy way to see the order of magnitude of numbers, there is a shorthand for them. We simply write $\log x$, without any base, to stand for $\log_{10} x$. Base 10 logarithms are also called *common logarithms*.

Note: Calculators have only two logarithm keys, one for base e logarithms (natural logarithms) and one for base 10 logarithms (common logarithms). Although there are formulas for expressing other logarithms in terms of either of these two, the IMP curriculum will not explore these formulas.

Key Questions

What do you suppose the special base is from *Find That Base!*?

How can you write 2 itself as a power of e?

How can this general form be simplified using e?

Does anyone recognize the number 0.693 from earlier in the unit?

How can you find the exact proportionality constant?

What do you call the exact solution to the equation $e^c = 2$?

California Population with e's

Intent

This activity has students find an exponential function of the form $y = k \cdot e^{cx}$ that goes through two given points.

Mathematics

In this activity, students reexamine the function they found in *California and Exponents* and express it in the form $y = k \cdot e^{cx}$.

Progression

California Population with e's asks students to find a function $y = k \cdot e^{cx}$ that passes through the two points used earlier in *California and Exponents*. It then asks students to express the value they get for c in terms of a natural logarithm and to relate both coefficients in this activity to those in the earlier activity.

Approximate Time

25 minutes for activity (at home or in class)

10 minutes for discussion

Classroom Organization

Individuals, followed by whole-class discussion

Doing the Activity

Students should have their work from *California and Exponents* for use in Question 3 of this activity.

Discussing and Debriefing the Activity

Let a volunteer present Question 1. As with their work on *California and Exponents,* students may have a variety of approaches for getting c once they find k. They will probably work from the equation $380{,}000 = 92{,}600 \cdot e^{c10}$ and simplify this (as in the earlier activity) to get $e^{10c} \approx 4.10$.

Students might solve this equation using guess-and-check, or they might apply the definition of natural logarithms to see that $10c$ must be equal to $\ln 4.10$. Even if no one expressed c as a natural logarithm for Question 1, they should have done so for Question 2, getting $c = (\ln 4.10)/10$ (which is approximately 0.141).

Finally, students should see that their value for k here is the same as their value for a in the earlier activity, but they may have more difficulty relating c to b. As a hint, bring out that e^{cx} must be equal to b^x, which means that e^c must be equal to b. In other words, c must be equal to $\ln b$. You can have students verify this numerically, using their previous value of approximately 1.15 for b. That is, they should see that $\ln 1.15$ is approximately equal to 0.141. (This works out more precisely if the more accurate value of 1.152 is used for b.)

Supplemental Activities

Transcendental Numbers (extension) asks students to research transcendental numbers, which include e and π.

Dr. Doubleday's Base (reinforcement) and *Investigating Constants* (extension) reinforce and then extend the work done with the proportionality property in *Slippery Slopes*.

Back to the Data

Intent

In this section, students solve the central unit problem.

Mathematics

The activities in *Back to the Data* provide an opportunity for students to synthesize and summarize what they have learned in this unit. Those topics include linear functions, slope, exponential functions, derivatives, average and instantaneous rate of change, the proportional relationship between exponential functions and their derivatives, logarithms, and fitting linear and exponential functions to data.

Progression

The centerpiece of this set of activities is *Return to A Crowded Place*, where students find an exponential function to approximately fit the population data from the unit problem and use that function to answer the question. *Tweaking the Function* helps to prepare students for that activity through an exploration of how changes in an exponential function affect the shape of its graph.

The remaining activities consist of the unit assessments and preparation of the unit portfolio.

Tweaking the Function
Beginning Portfolios—Part I
Return to A Crowded Place
Beginning Portfolios—Part II
Small World, Isn't It? Portfolio

Tweaking the Function

Intent

In this activity, students explore how to make changes to an exponential function to modify its graph.

Mathematics

Students learned in *California with e's* to find an exponential equation whose graph passes through two given points. For the unit problem, however, they will need to find a function that approximates a larger array of data. With fitting it to two points as a starting point, they need to have some intuition about how to modify the function in order to adjust it to better represent the entire set of data. In this activity, students explore in order to develop that understanding.

Progression

Tweaking the Function is a free-for-all exploration. Students use the graphing calculator to explore how changing various components of an exponential function changes its graph.

Approximate Time

40 minutes for activity

Classroom Organization

Small groups

Doing the Activity

In *Return to A Crowded Place,* students return to the data from the unit's opening activity and try to find a function that fits those data. In *Tweaking the Function,* which is preparation for that work, they simply explore the effect of various changes to the basic exponential function $y = e^x$.

Emphasize to students that this activity has no specific goal other than building their intuition—it is a "free-for-all" like others they have done.

There are no discussion notes for this activity. You might let groups share their insights toward the end of class, or you might simply have each group use its own results when it works on *Return to A Crowded Place.*

Beginning Portfolios—Part I

Intent

In this activity, students summarize ideas about linear and exponential growth.

Mathematics

This activity is a chance for students to reflect on their work on using linear and exponential functions to represent growth. It asks students to compare linear and exponential functions, discussing how each function represents rates of growth, how each represents initial values, and what kinds of situations each best fits.

Progression

This activity will be part of the unit portfolio. Allow students to share ideas following this activity and to modify their writing before including it in the portfolio.

Approximate Time

25 minutes for activity (at home or in class)

5 to 10 minutes for discussion

Classroom Organization

Individuals, followed by whole-class discussion

Doing the Activity

Tell students that this activity will be part of their unit portfolio.

Discussing and Debriefing the Activity

Let volunteers share their ideas, explaining the roles of the specific constants in each form. Students should see that the variable a in the expression $a + bx$ and the variable k in the expression $k \cdot e^{cx}$ play similar roles, giving the value of the expression when x is 0. They should also see that b and c play similar roles, governing the rate of growth.

In the discussion of appropriate contexts for each type of function, students should see that linear functions are appropriate when the amount of growth depends only on the change in x, while exponential functions are appropriate when the rate at which something grows depends on how big it already is. (This is essentially the principle that the derivatives of exponential functions are proportional to the functions.)

Return to *A Crowded Place*

Intent

In this activity, students solve the central unit problem.

Mathematics

Solving the unit problem requires that students search for a function to fit their population data. Knowing now that exponential functions are best suited to modeling population growth, they will need to begin from an exponential function and adjust it to best fit the set of data.

Progression

Students plot the population data from the central unit problem on their graphing calculators and then find an exponential function that approximates the data. They make a prediction to answer the unit problem, then write about whether they think the population data are truly exponential or will continue to be so.

Approximate Time

35 to 40 minutes for activity

40 minutes for presentations and discussion

Classroom Organization

Small groups, followed by presentations from each group and whole-class discussion

Doing the Activity

Ask, **What is the central unit problem?** Students will probably remember the list of data but may not recall the specific task. If not, have them reread the opening activity, *A Crowded Place*. As stated there, the central question is this:

If this pattern of data continued, how long do you think it would take until we were all squashed up against one another?

If necessary, remind students that they developed a specific definition of what it meant to be "squashed up against one another" and used that definition to refine the central question to this:

How long will it take until the population reaches $1.6 \cdot 10^{15}$ people?

Once this revised question has been restated, ask students how they might answer it. Lead them to the idea of finding an exponential function, or a variation of one, that comes close to fitting the data and then using that function to answer the question.

With that introduction, have groups begin work on *Return to "A Crowded Place".* You may want to urge them to use any insights they gained from *Tweaking the Function.* But you should also warn them that they may have great difficulty getting a function that comes near all of their data points.

Give each group a transparency to use in preparing a presentation of their solution.

Further Suggestions

You may want to suggest to students that they start by picking two of the data points. They can use the approach of *California Population with e's* to get a function of the form $y = k \cdot e^{cx}$ that goes through their two chosen points.

Once they have that function as a first guess, they can apply ideas from *Tweaking the Function,* varying k and c in an attempt to get a better fit for the other points.

Discussing and Debriefing the Activity

For this activity, each group is likely to come up with a different result, so have all groups make brief presentations. Have a student from each group give the function they used for Question 3, explain how they decided on that function, and give the answer they found for when the population reaches $1.6 \cdot 10^{15}$ people.

It may work best to discuss Question 4 as a whole class after all of the individual presentations. The class will probably conclude that the population data set in the activity is not strictly exponential, and they should certainly agree that population growth will not continue to be exponential for the long-term future. Have students give reasons the relative growth rate will not be constant.

Key Question

What is the central unit problem?

Beginning Portfolios—Part II

Intent

This activity continues students' work in reflecting on the unit and preparing their portfolios.

Mathematics

In this activity, students reflect upon what they have learned about compound interest and about finding equations of straight lines.

Progression

Beginning Portfolios—Part II will be included in the unit portfolio. As with the previous part, allow students to share ideas and to modify their writing before attaching this to their portfolios.

Approximate Time

20 minutes for activity (at home or in class)

5 to 10 minutes for discussion

Classroom Organization

Individuals, followed by whole-class discussion

Doing the Activity

Tell students that this activity will be part of their unit portfolio.

Discussing and Debriefing the Activity

Have a couple of students share their ideas on each of the two parts of the activity. You can use this discussion as an opportunity to review some of the fundamental ideas of the unit.

Small World, Isn't It? Portfolio

Intent
In this activity, students prepare the unit portfolio.

Mathematics
The portfolio cover letter will include discussion of the main mathematical ideas from the unit, how they were developed, and how they were used to solve the central unit problem.

Progression
Students will have begun their portfolio work in *Beginning Portfolios—Part I* and *Beginning Portfolios—Part II*, so their main tasks in this activity are to write their cover letters and discuss their personal growth.

Approximate Time
30 minutes for activity (at home or in class)

15 minutes for discussion

Classroom Organization
Individuals, followed by whole-class discussion

Doing the Activity
Students work independently writing their cover letter and selecting portfolio activities that reflect the mathematical ideas learned in the unit.

Discussing and Debriefing the Activity
Let volunteers share their portfolio cover letters as a way to start a discussion to summarize the unit.

Blackline Masters

How Many More People?

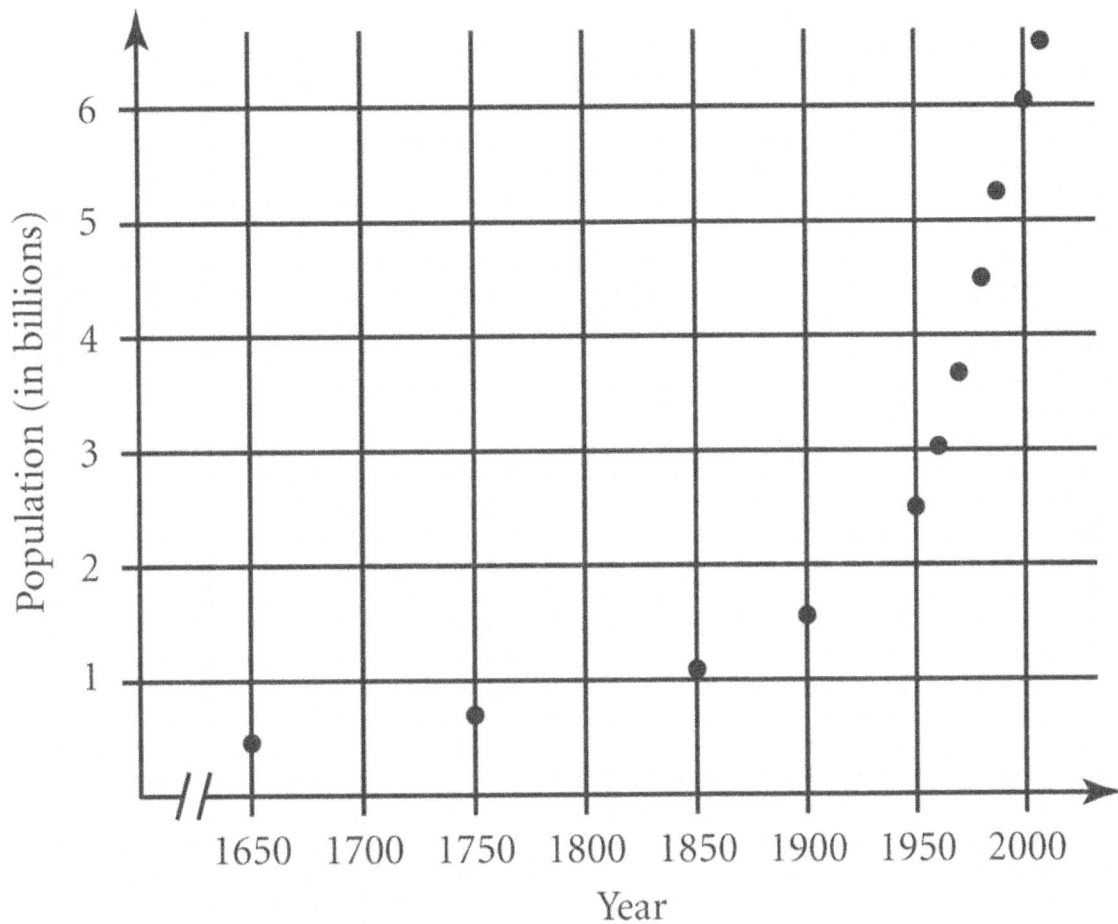

Growing Up

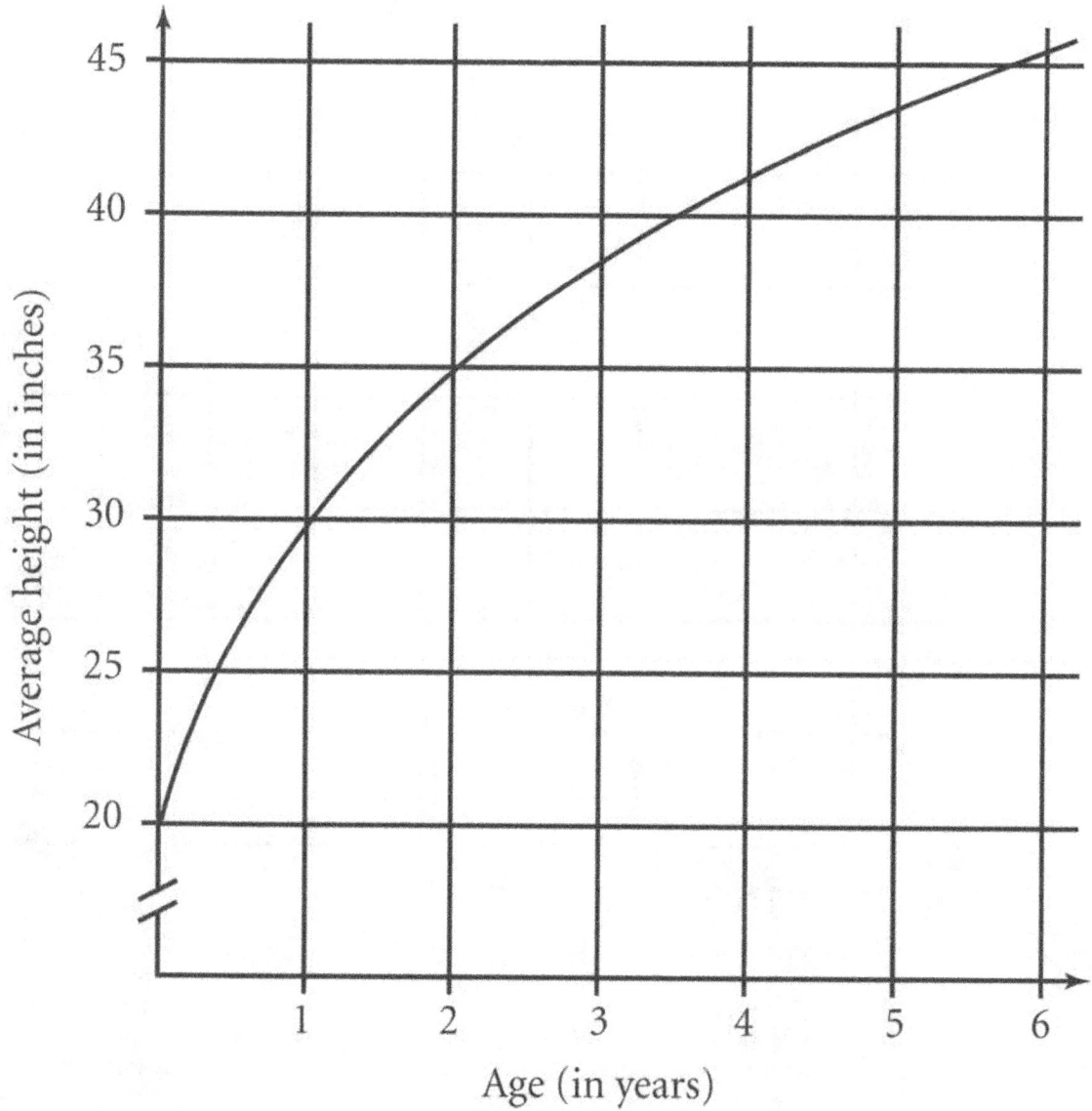

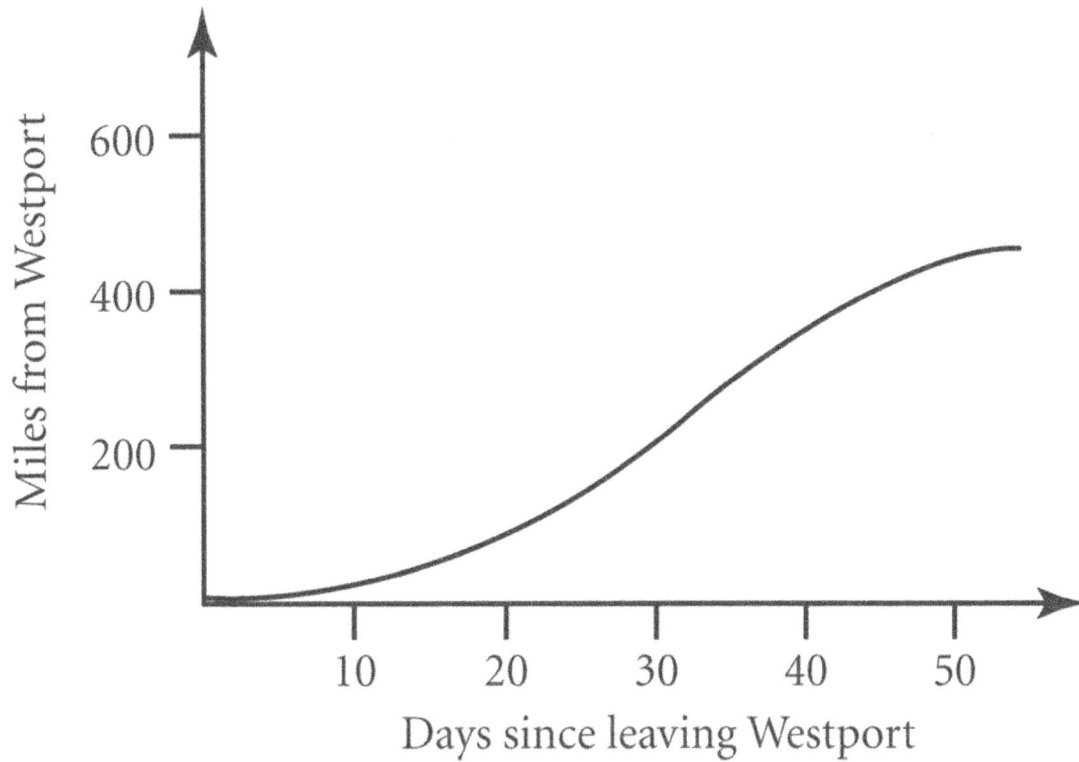

Traveling Time

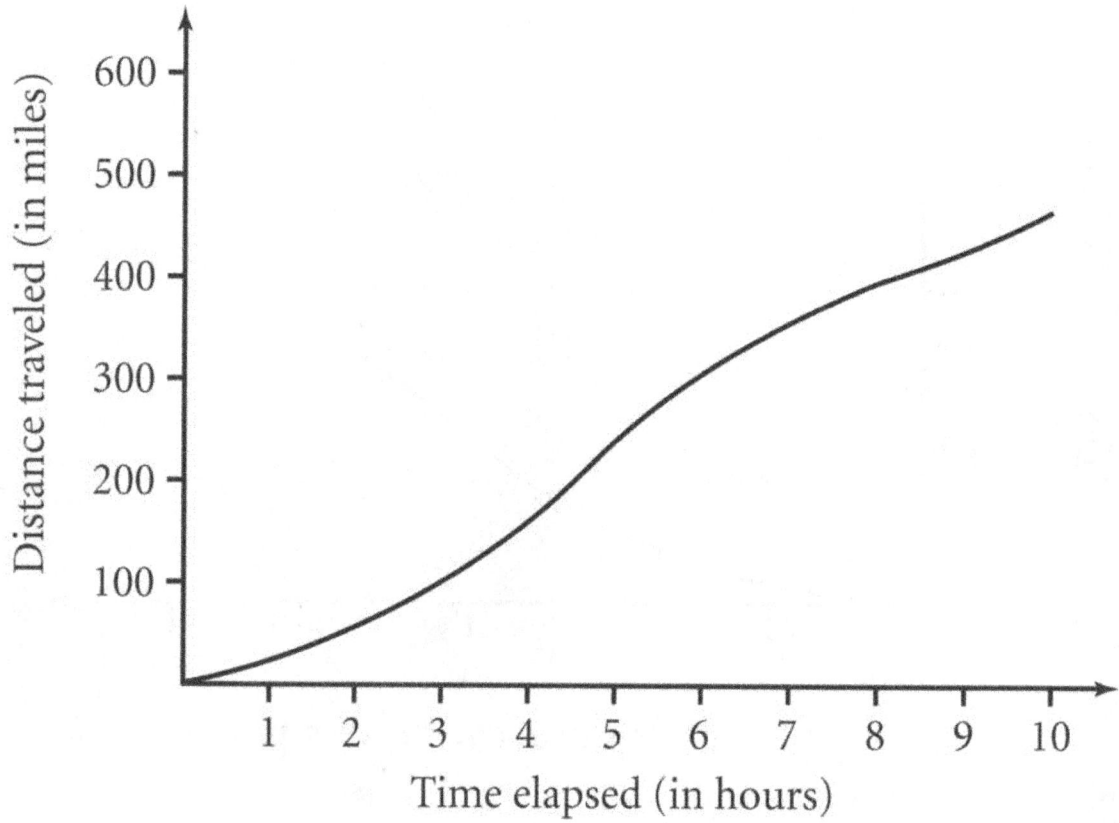

Comparative Growth

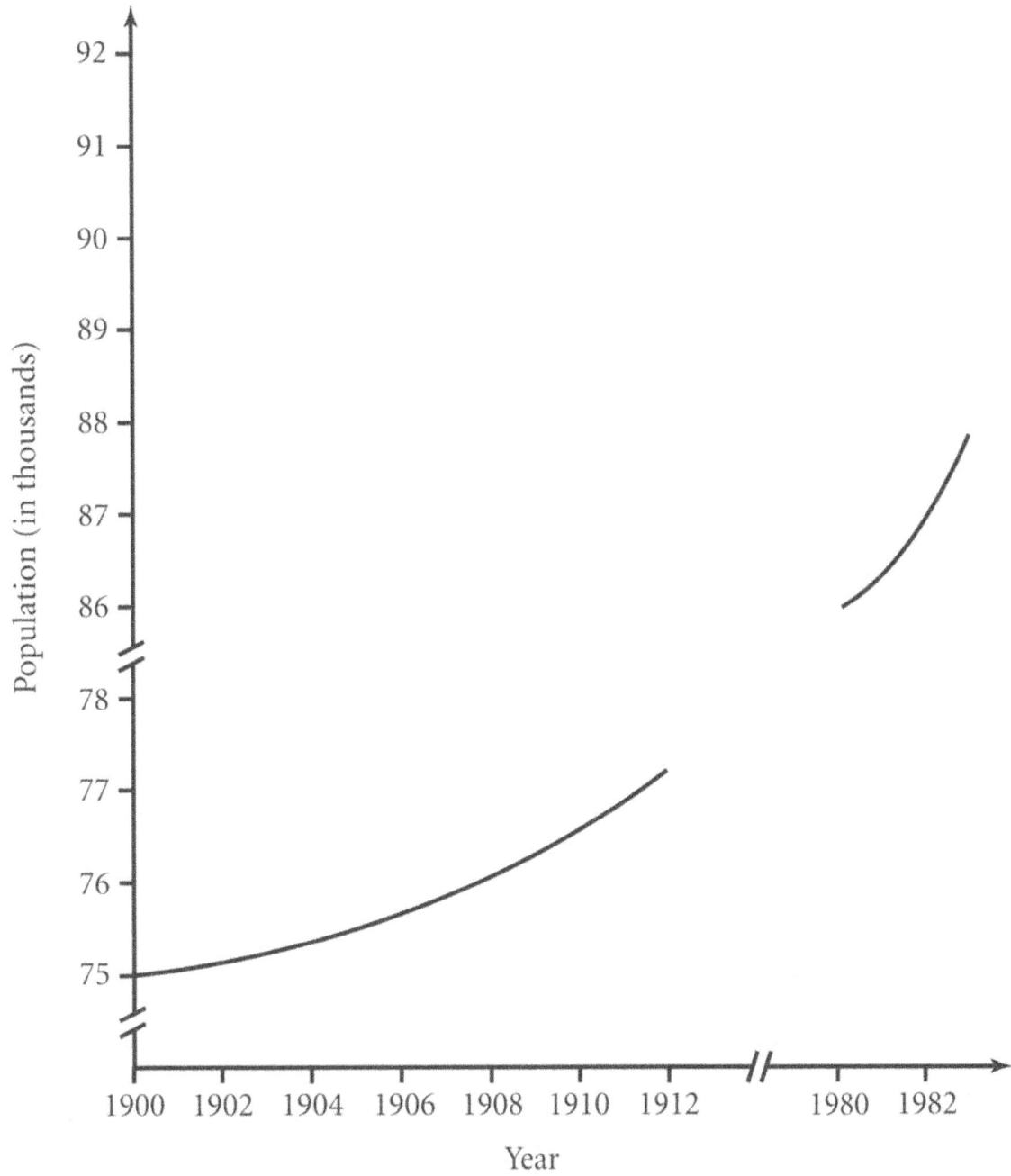

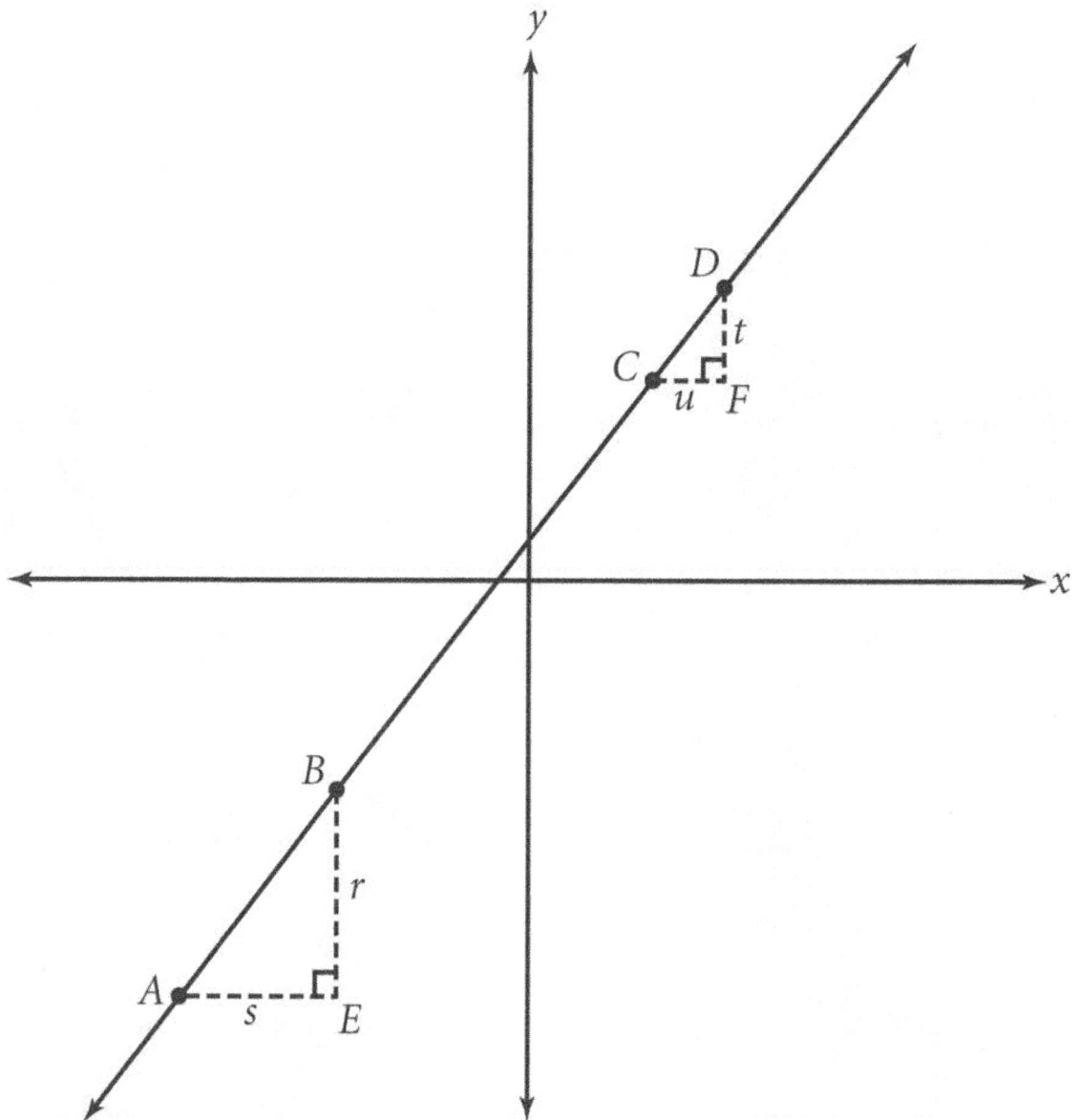

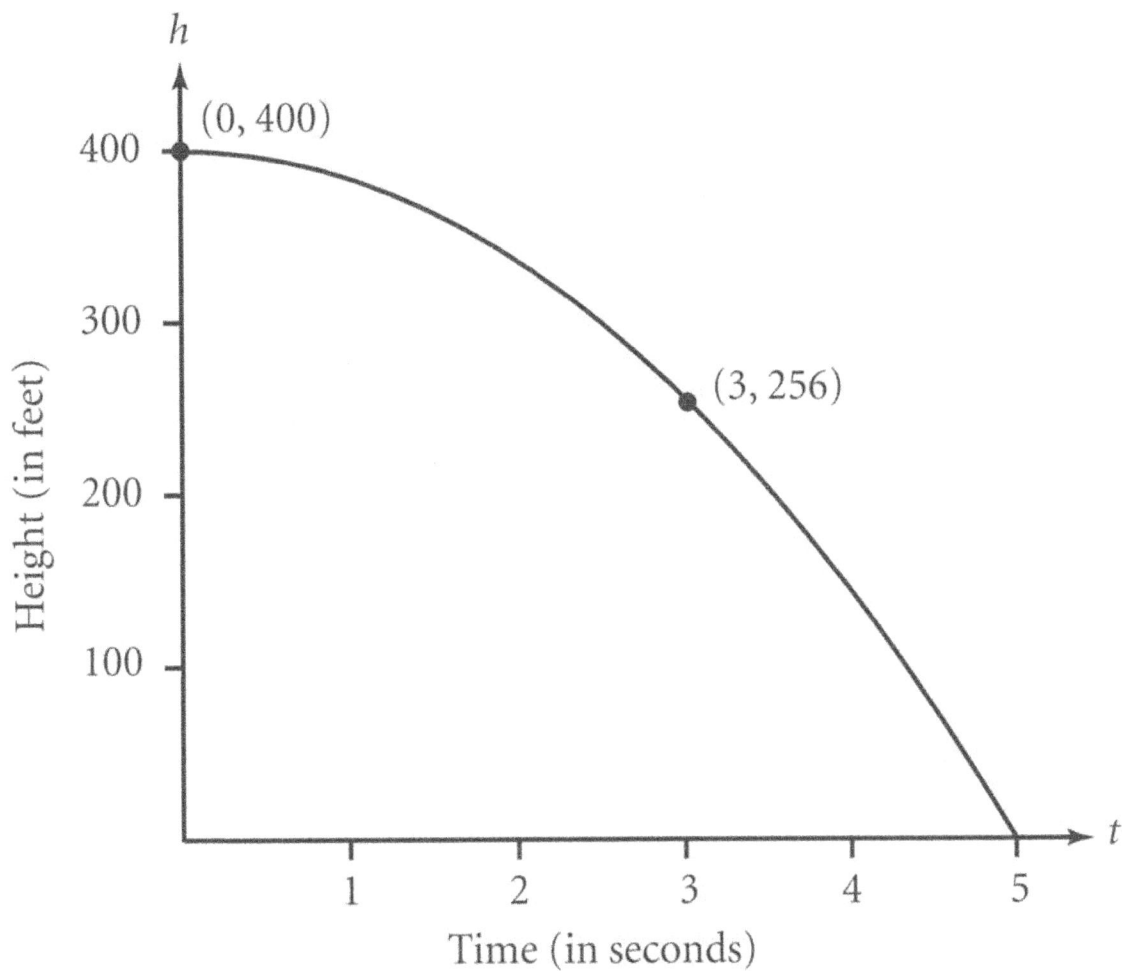

The Growth of the Oil Slick

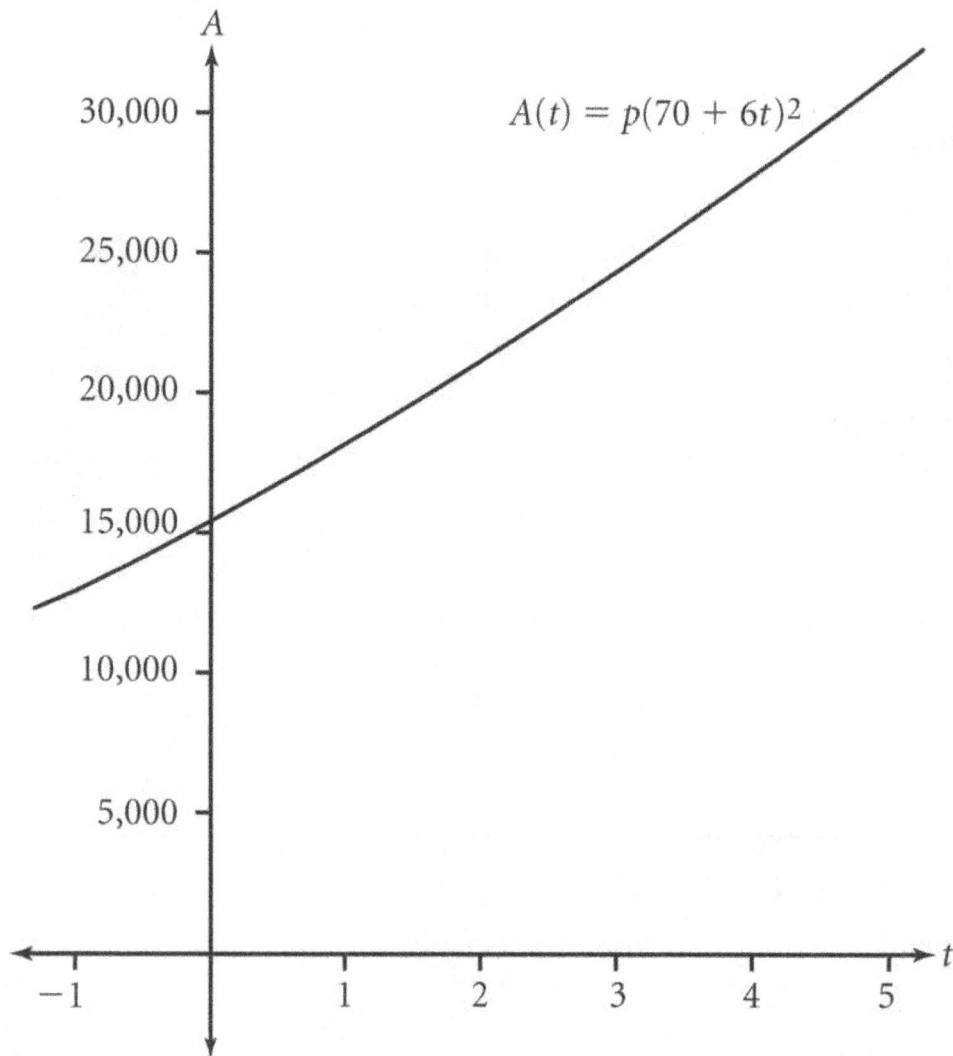

$$A(t) = p(70 + 6t)^2$$

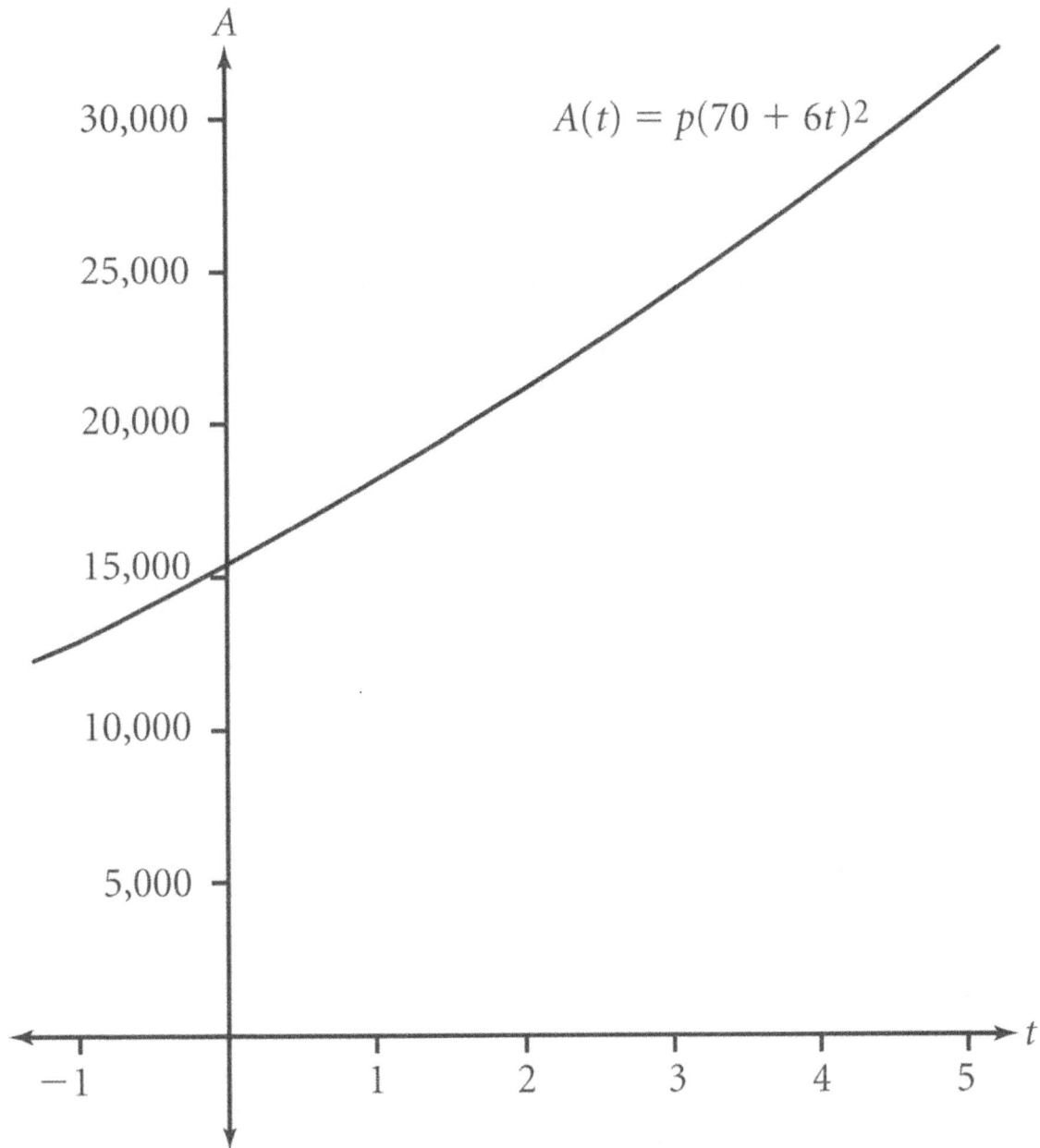

$A(t) = p(70 + 6t)^2$

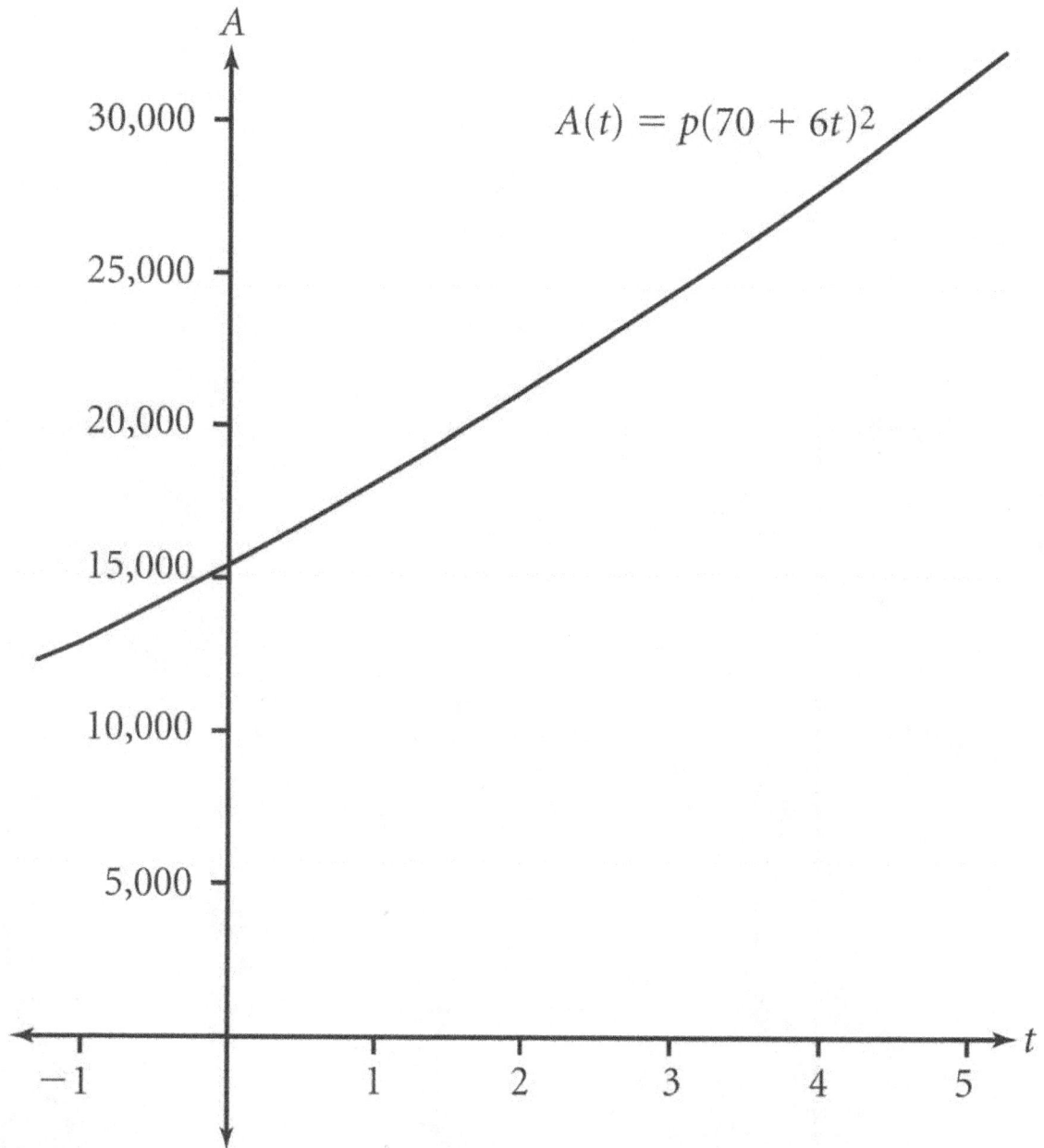

$$A(t) = p(70 + 6t)^2$$

¼-Inch Graph Paper

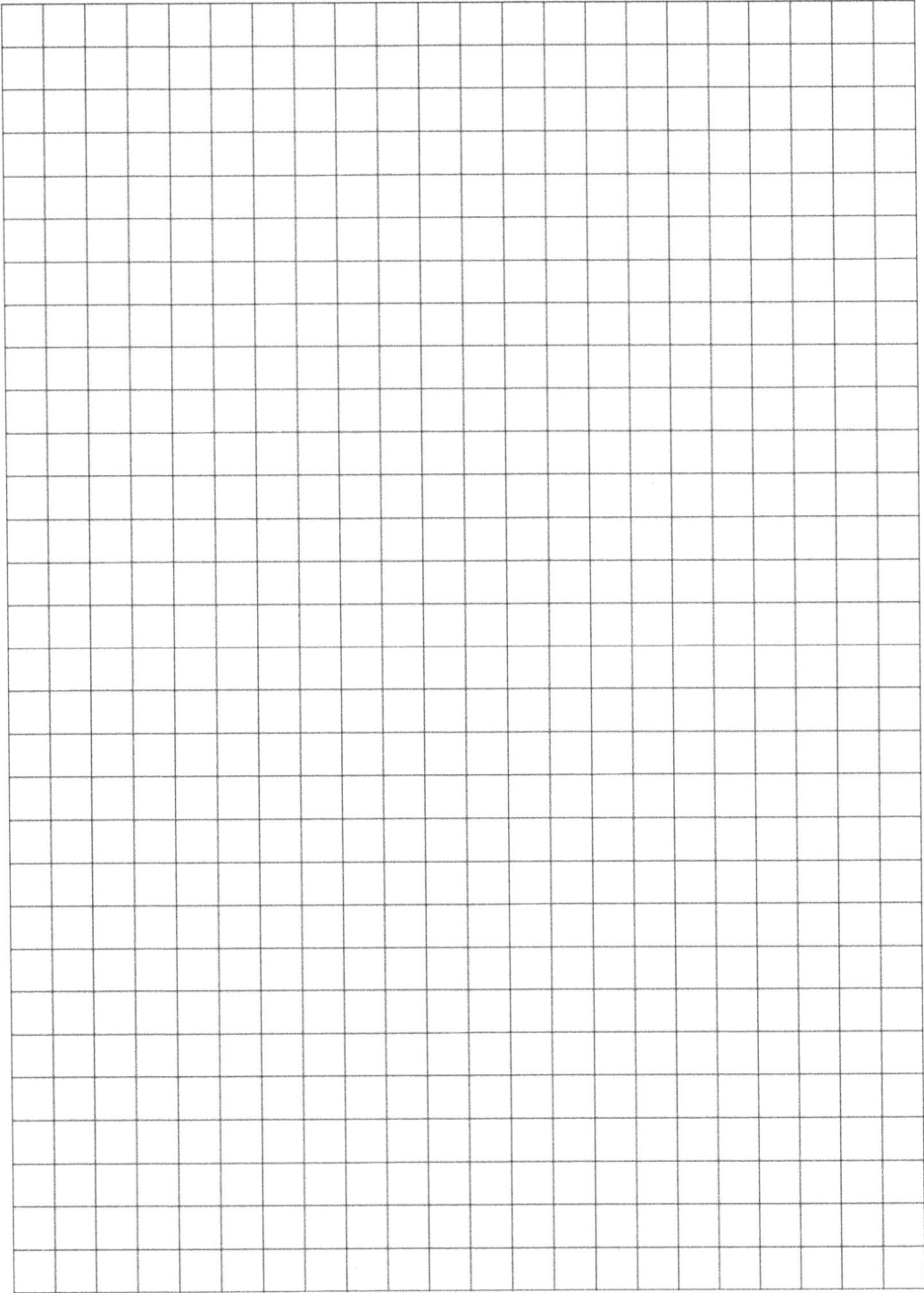

© 2011 Interactive Mathematics Program

1-Centimeter Graph Paper

1-Inch Graph Paper

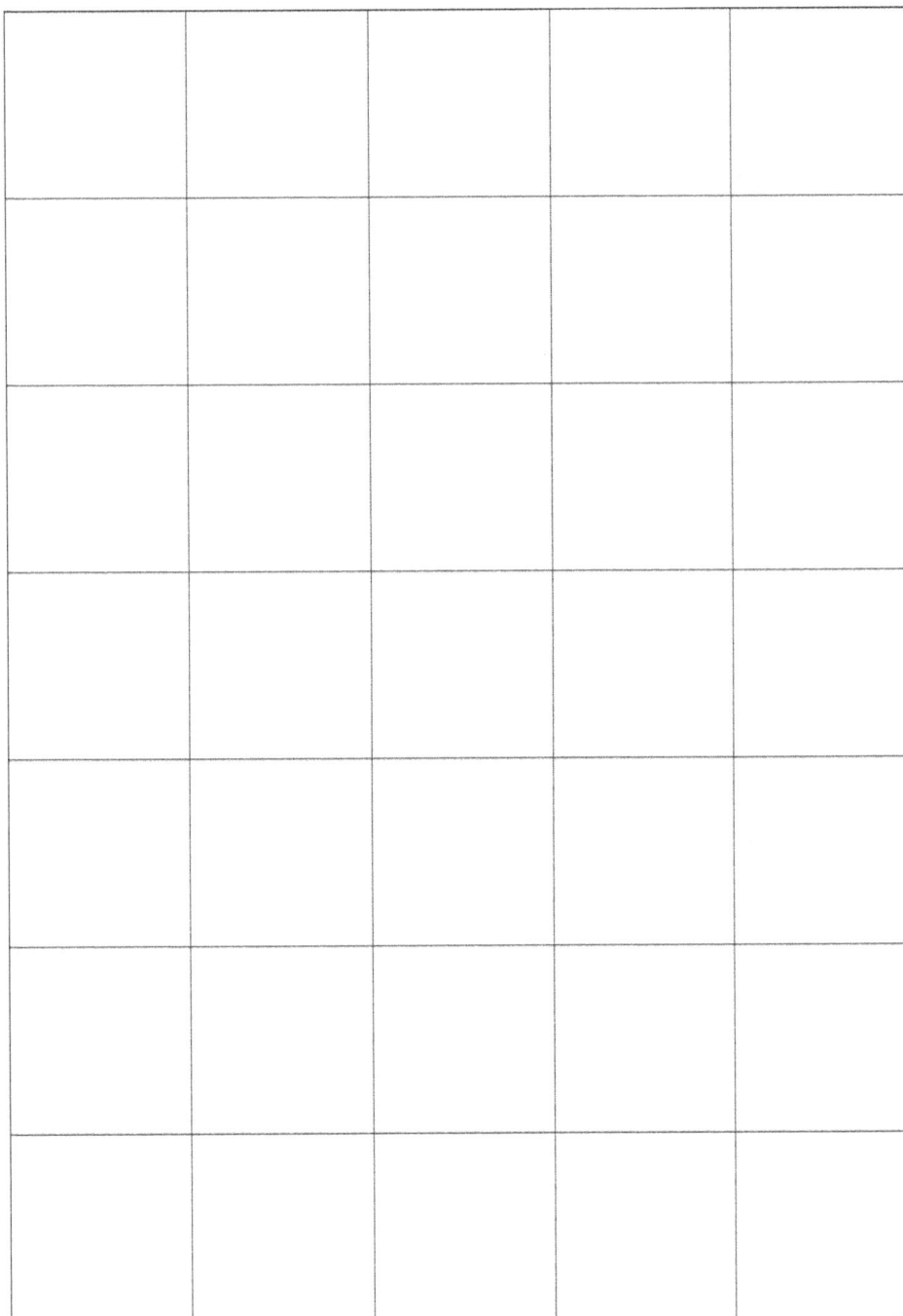

Assessments

In-Class Assessment

1. Bill bought a used car. After he'd owned the car for 10 weeks, he noticed that the car's odometer showed that it had gone a total of 43,852 miles. (Of course, that included the mileage on the car when Bill bought it.) It's now 30 weeks since Bill bought the car, and the odometer shows 49,651 miles.

   Assume that Bill drives about the same amount each week.

   a. Develop a formula that will estimate the odometer reading after Bill has owned the car for *N* weeks.

   b. About how many miles had the car gone when Bill bought it? Explain your reasoning.

   c. If you graphed the function defined by your formula, what would be the slope of that graph? Explain your reasoning.

2. An object is thrown into the air off the roof of a building so that its height (in feet) after *t* seconds is given by the formula $h(t) = 160 + 20t - 16t^2$.

   a. What is $h'(2)$? That is, what is the derivative of this function at $t = 2$?

   b. What does the answer to Question 2a mean in terms of the problem situation?

Take-Home Assessment

1. Red Rock Reservoir

   The Red Rock Water Works keeps careful watch over the water level of Red Rock Reservoir. The graph below shows the amount of water in the reservoir over the course of a year. Use this graph to answer Questions 1a through 1c. Estimate your answers as needed.

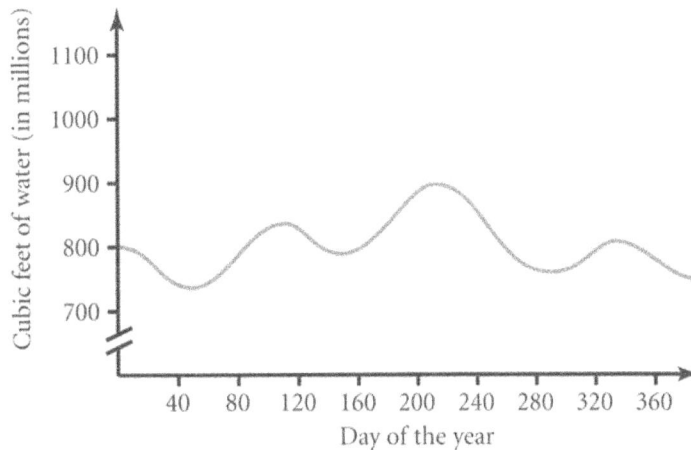

   Reservoir Volume During a Year

   a. What is the water level on Day 120? On Day 210?

   b. Approximate the value of the derivative of this function at Day 60. What does this value mean in terms of the problem situation?

   c. During what periods is the water level rising? During what periods is it dropping? During what periods is it staying the same? What do your answers tell you about the derivative for each of these time periods?

2. Populations and Exponential Functions

   a. What is the difference between absolute growth rate and proportional growth rate? Give an example of an equation modeling each kind of growth, and describe a real situation involving each kind of growth.

   b. Discuss why exponential functions are often useful for representing the size of a population over time.

IMP Year 3
First Semester Assessment

I.   Once Upon a Time . . .

Imagine that you are Madie or Clyde. You've grown old and are telling your grandchildren the story of the orchard hideout. You've described the arrangement of trees in the original orchard (with a radius of 50 units) and told them the basic facts that you knew at the start.

- The circumference of the newly planted trees
- The fixed amount by which the cross-sectional area of the trees grew each year
- The distance between the centers of adjacent trees

Of course, the grandchildren have heard the story before, and they remember that it took about 11 years, 9 months for the center to become a true orchard hideout. What you want to do is impress them with how well you and your partner analyzed the problem back then.

Write a description of how the analysis worked. Don't get bogged down in the specific numbers, because you don't have pencil and paper handy, and the youngsters are more interested in the big ideas anyway!

II.   Road Building

The highway department is planning a road that will go through the town of Coldwater. The town of Hot Springs is 13 miles due south of Coldwater, and the town of Warm Rock is 18 miles due east of Hot Springs.

1. Sketch a diagram showing the relationship between the three towns. (Treat each town as a single point.)

The mayors of Hot Springs and Warm Rock both want this new road through Coldwater to go straight through their towns as well. Unfortunately, the highway department can afford to build only one road.

The road must go through Coldwater and must be straight, so a compromise route is needed. The mayors of Hot Springs and Warm Rock agree to support the project if this condition is met:
The distance from Hot Springs to the new road must be the same as the distance from Warm Rock to this road.

They also insist that the road should not be parallel to the route from Hot Springs to Warm Rock.

2. a. Add a dotted line to your sketch from Question 1 to show where the new road must go, and explain your reasoning.

b. Find the distance from Hot Springs to the road, to the nearest tenth of a mile, and explain your work.

III. Equation Time

Solve this system of equations, and explain your work.

$$7r + 6s = 6$$
$$5r - 4s = 25$$

IV. The Third Dimension

This system of linear constraints in three variables defines a feasible region.

I	$2x + y + z \leq 20$
II	$3x + 4z \leq 13$
III	$y + 2z \leq 9$
IV	$x \leq 9$
V	$y \leq 6$
VI	$x \geq 0$
VII	$y \geq 0$
VIII	$z \geq 0$

Give *a general outline* of how to find the point in this feasible region where the function $2x + y + z$ has its maximum, and explain the geometric reasoning behind your method.

V. Solving with Matrices

Consider this system of linear equations.

$$3a + 2b - c + d = 1$$
$$2a - b + 4c + 2d = -2$$
$$-4a + 3c - 3d = -6$$
$$a + b + c + d = 3$$

1. Write a matrix equation that is equivalent to this system.

2. Solve the system using matrices on a graphing calculator, and show your solution.

3. Discuss the relationship between the matrices and the equations and the properties of matrices that allow you to use them to solve systems of linear equations.

IMP Year 3
Second Semester Assessment

I. Spilt Milk

You've automated your dairy farm so that all the cows are milked by milking machines, and the milk all flows into one giant cone-shaped container. At the start of milking time, the container is empty, and as the milk flows in, the level in the container rises. Milking starts at 5:00 a.m. and continues through the day. (The cows are not all milked at the same time.)

After studying your cows and using some geometry, you've figured out that at t minutes after 5:00 a.m., the milk in the container will have risen to a level of $\sqrt[3]{2000t}$ centimeters.

1. During the hour from 7:00 a.m. until 8:00 a.m., what is the *average* rise per minute in the height of the milk? (Give your answer to the nearest 0.001 cm/min.)

2. At what rate is the milk level rising at 8:00 a.m.? (Again, give your answer to the nearest 0.001 cm/min.)

3. At what time of day will the milk level reach 100 centimeters?

II. Darts

Consider a square dartboard with a circle inscribed in the square, as shown here.

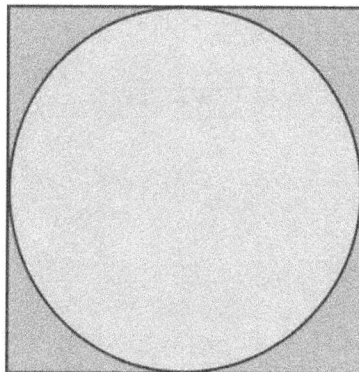

Suppose that according to the rules, if your dart lands inside the circle, you win, and if the dart lands outside the circle, you lose. Assume that you always hit the dartboard and that each point of the square is equally likely to be hit.

1. If you throw one dart, what is your probability of winning? Explain your answer, giving the probability to the nearest hundredth.

2. Suppose you throw seven darts. What is the probability that you will win at least four times? Explain what method you use to find the answer and why the method works. Again, give the probability to the nearest hundredth.

III. Ferris Wheel Fence, Revisited

It's time to look back at the problem of the fence around the amusement park, from *High Dive*.

As you may recall, Al and Betty are riding on a Ferris wheel. This Ferris wheel has a radius of 30 feet, and its center is 35 feet above ground level. There is a 25-foot-high fence around the amusement park, but once you get above the fence, there is a wonderful view.

What percentage of the time are Al and Betty above the level of the fence?

IV. Opposite Angles

You have learned these formulas involving trigonometric functions:

$$\cos(-\theta) = \cos\theta$$
$$\sin(-\theta) = -\sin\theta$$

Explain each of these formulas in several ways:

- In terms of the Ferris wheel
- In terms of the graphs of the sine and cosine functions
- Using numerical examples

You can use these graphs of sine and cosine in your explanation:

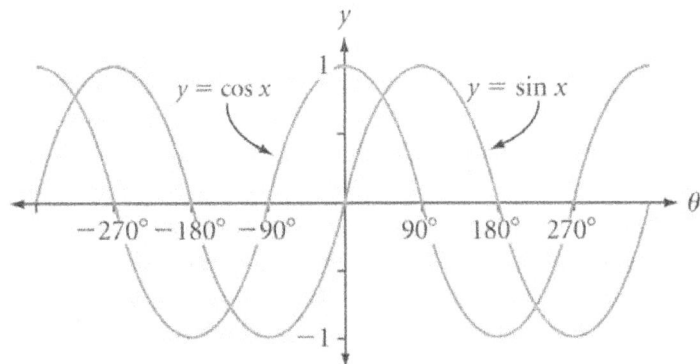

You can also use this diagram to represent a Ferris wheel:

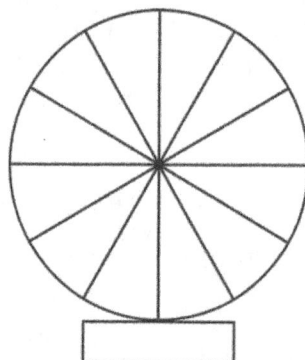

Small World, Isn't It? Calculator Guide for the TI-83/84 Family of Calculators

The primary calculator topics in *Small World, Isn't It?* are slope, rate of change, and derivative in both numerical and graphical contexts. As students explore rate-of-change problems, the calculator's graphing capability allows them to formulate and test algebraic conjectures hand-in-hand with their graphic representations. Exponents, logarithms, and scientific notation will all play a part in the unit, but shouldn't require new calculator skills.

A Crowded Place: As students begin their initial work with the unit problem, they may reach for their graphing calculators to plot the data pairs. This may be a great way for some students to begin to investigate patterns in this data set, even if they have little success. As directed in the activity, they should make note of any difficulties they encounter. Consult the Calculator Note "Plotting Points," found in *Calculator Basics* in the Year 3 general resources, for a brief introduction to plotting points on the graphing calculator.

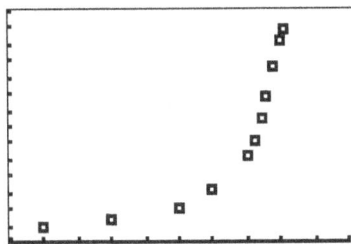

This unit problem involves very large numbers. Thus, students will be challenged throughout the unit to keep track of these numbers and to correctly read scientific notation on the graphing calculator. They have studied scientific notation in the Year 2 unit *All About Alice*, so these ideas should be familiar.

Note the list of data illustrated on the screen, especially the entries in **L2** for 1750 and 1850. This type of difference is usually sufficient to remind students about the advantages of scientific notation.

How Many of Us Can Fit? and *How Many More People?:* The possible ways to use the calculator are similar to those from *A Crowded Place.* The same issues about plotting points and scientific notation may come up during work on these activities.

Story Sketches: During the discussion of this activity, you may find it helpful to employ certain capabilities of the graphing calculator while discussing the differences between continuous, discrete, and step functions. The following comments are for your reference; the focus of the day's discussion should be on the properties of straight-line graphs.

Some students are likely to draw a continuous graph as shown here for Tyler's accumulated savings. This is done simply by entering **.5x + 2** in the $\boxed{\text{Y=}}$ editor and setting the appropriate viewing window.

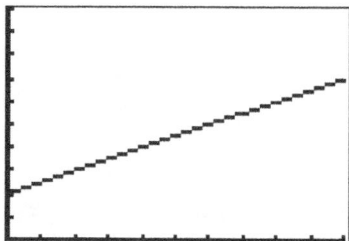

A discrete graph can *occasionally* be modeled well on the graphing calculator. The rule for Tyler will not. (You will usually have better luck with discrete-looking graphs when the *x*-coefficient is larger, such as values of 3 or more.)

To see if a discrete graph will appear on the calculator, enter the function in the $\boxed{\text{Y=}}$ editor. In this example, use the expression **3x**. Then press $\boxed{\text{MODE}}$, highlight **Dot** instead of **Connected**, press $\boxed{\text{ZOOM}}$, and select **6:ZStandard**. The graph will appear, but tracing will show that the calculator has plotted every *x*-value in increments of approximately 0.21. Change this by pressing $\boxed{\text{ZOOM}}$, highlighting **8:ZInteger**, and pressing $\boxed{\text{ENTER}}$ to return to the graphing screen, and pressing $\boxed{\text{ENTER}}$ again. Then press $\boxed{\text{TRACE}}$. Tracing will show *x*-values in increments of 1. This method may not work consistently in the way you wish, but it should be helpful during classroom discussions.

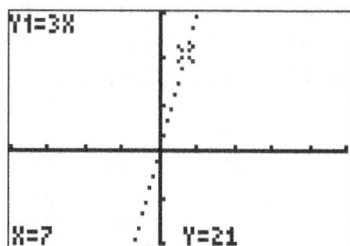

Tyler's accumulated savings may be best described as a step function, and you can model step functions on the calculator. The command **int(** is the greatest integer function and will return the largest integer less than or equal to the value inside the parentheses. (The greatest integer function and its standard notation are introduced formally in IMP Year 4.)

Enter **.5int(X)+2** in the $\boxed{\text{Y=}}$ editor. Find **int** by pressing $\boxed{\text{MATH}}$, highlighting **NUM**, highlighting **5:int(**, and pressing $\boxed{\text{ENTER}}$. Leave the calculator graphing mode set to **Dot** instead of **Connected**. Set an appropriate window viewing and then press $\boxed{\text{GRAPH}}$.

```
Y1=.5int(X)+2
                    .___
               ___
          _>|<_
     ___
__
X=5              Y=4.5
```

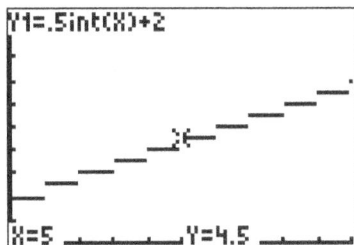

What a Mess!: This activity is conducive to using the graphing calculator. Students should recognize that they want the radius of the oil slick to be equal to the radius of the clean-water circle. They can set up equations for the two radii, being careful to use the same meaning for t in the two equations. (For instance, if they use $t = 0$ to represent the time when the cleanup began, the equations would be equivalent to $r = 112 + 6t$ and $r = 10t$.) Students can then use either the graphing capabilities or function tables on their calculators to solve this pair of equations. See "Solving Systems by Graphing" in the *Meadows or Malls?* Calculator Notes for further instructions.

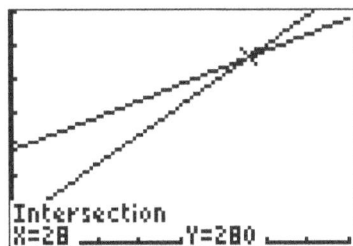

```
Intersection
X=28            Y=280
```

Points, Slopes, and Equations: During the discussion of this activity, you might ask students to verify if a particular ordered pair fits the equation. Students can test the ordered pair on the graphing calculator by entering the equation into the $\boxed{Y=}$ editor. They may then check the graph or table to see if the point fits.

Another method available is to perform substitution using function notation. If the expression **2+5(x−3)** is entered for **Y₁**, you might simply enter **Y₁(3)** in the home screen to evaluate **Y₁** for an x-value of 3. (To get **Y₁**, press $\boxed{VARS}$, highlight **Y-VARS**, and then select **1:Function**.)

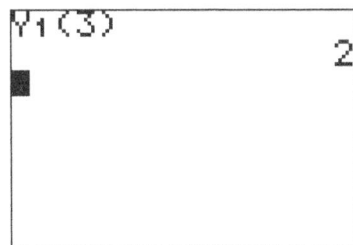

```
Y1(3)
                    2
```

Return of the Rescue: During the discussion of this activity, students should graph the function $h(t) = 400 - 16t^2$. When introducing the term *secant line*, you may wish to demonstrate a line connecting two points on a graph using the calculator.

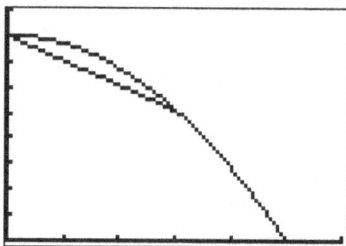

Once the graph is displayed on the calculator, press 2ND [QUIT] to return to the home screen. Next, press 2ND [DRAW] and select **2:Line(**. Complete the command by entering the coordinates of two points, each separated by commas, and pressing ENTER. The screen shown above used the command **Line(0,400,3,256)**.

Photo Finish: Students are likely to use the graphing calculator with this activity. If students wish to substitute a particular value for *t* in the formula, once again they can use function notation on their calculator. See the description under *Points, Lines, and Equations* for details on how to do this.

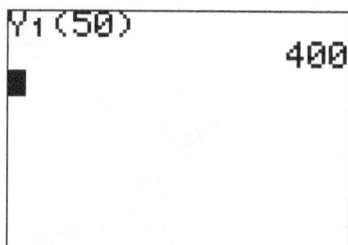

Students might solve Question 2 by calculating the average speed during the last tenth of a second. The calculation illustrated in this screen uses the function $m(t) = 0.1t^2 + 3t$. On your calculator, first enter this function as **Y₁**, and then duplicate what you see in this screen to calculate the average speed from $t = 49.9$ to $t = 50$. (Refer to the description under *Points, Lines, and Equations* for instructions on how to enter **Y₁** on the home screen.)

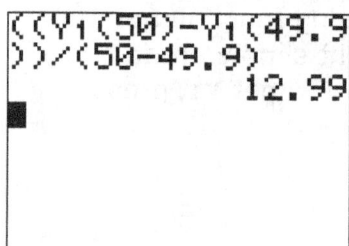

The supplemental problem *Speedy's Speed by Algebra* generalizes this approach and can also be done in conjunction with the graphing calculator.

ZOOOOOOOOM: This activity centers around the graphing capabilities of the calculator. Students should be encouraged to use only the TRACE and ZOOM features in completing *ZOOOOOOOOM*.

Begin by entering Speedy's rule in the Y= editor. Next, determine a reasonable window to view the entire 400-meter race. The second screen below shows one possibility. Press GRAPH to view the function graph. Then press TRACE to move the cursor to a point near 400 meters (the end of the race).

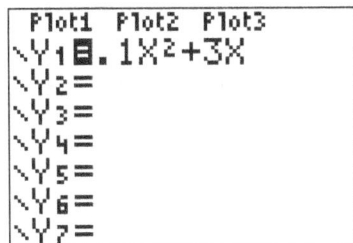

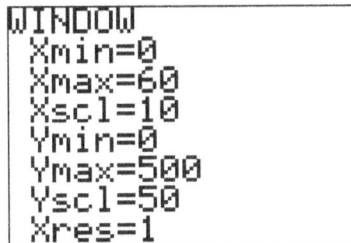

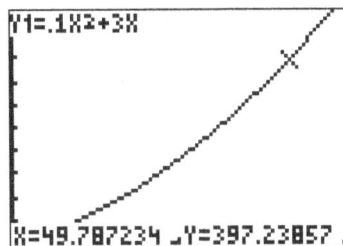

Now, press ZOOM 2 ENTER ENTER to zoom in on the graph. Try to get an ordered pair with a *y*-coordinate as close as possible (if not exactly) to 400 meters.

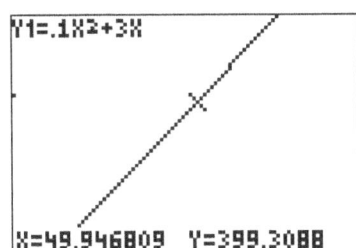

Students can repeat this process until they are satisfied with their estimates. Using points found by tracing after having zoomed in on the graph should help to emphasize the geometric meanings of slope and rate of change.

In the discussion following *ZOOOOOOOOM*, students consider the graph of a line tangent to a curve at some point, and the way it relates to instantaneous rate of change.

The calculator will draw tangent lines; brief details are included here. It is neither expected nor recommended that students learn this calculator skill at this point in their investigation of instantaneous rates of change. However, you may find this helpful in demonstrating tangent lines, or possibly as an idea for an interested student to pursue later on.

Graph the function in a reasonable viewing window. This example uses Speedy's function from *Photo Finish, m(t) = 0.1t^2 + 3t*. Then go to the home screen (2ND [QUIT]). Enter the command **Tangent(Y1,50)** and press ENTER. (To find **Tangent**, press 2ND [DRAW].)

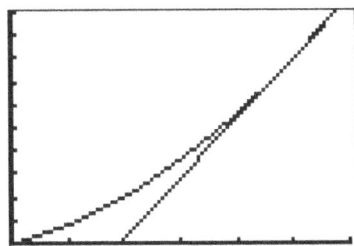

Speeds, Rates, and Derivatives: By this point in the unit, students may have identified the feature on their calculator that determines the derivative of a function at a single point. Although this seems like a nice shortcut, you may wish to consider what risks might be involved if students move away from calculating average rates of change by hand to the more abstract process of allowing the calculator to do it for them. The Calculator Note "Derivative at a Point" provides instructions on how to find derivatives on the calculator. (The numerical derivative can be used in several ways, including in programs and in the $\boxed{Y=}$ editor.)

Zooming Free-for-All: Question 3 of this activity may elicit the idea that some graphs "peak" or come to a point. The absolute value function does just this.

Have students graph the absolute value function by entering **abs(X)** in the $\boxed{Y=}$ editor. To do this, press $\boxed{MATH}$, highlight **NUM**, and press $\boxed{ENTER}$. It is a powerful learning moment for students when they zoom in repeatedly, recognizing that this function will never appear straight, while all the others they had looked at do. (*Note:* The calculator's **nDeriv** function incorrectly shows that the absolute value function has a derivative equal to 0 at $x = 0$.)

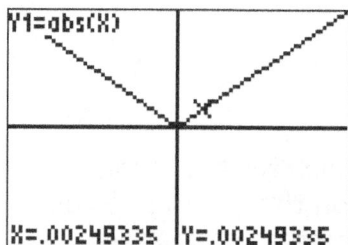

If students sketch an absolute value graph other than $y = |x|$, they may want to try drawing their graph on the calculator. You might start them off with a function defined by an expression like **–2abs(x–2)+5**, as shown here.

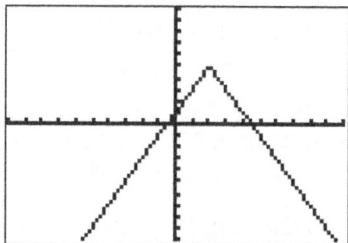

Some students may describe another type of function with peaks, such as **4abs(sin(x))**. Students will need to adjust the window settings to get the display shown here.

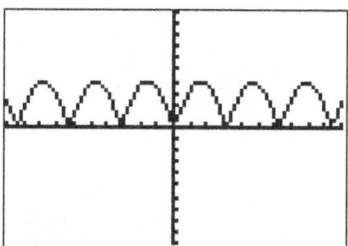

Around King Arthur's Table: As students make presentations of *POW 9: Around King Arthur's Table*, you may decide that it is appropriate to introduce the concept and notation for the greatest integer function. This concept and notation were used previously only in the Supplemental Activity *Integers Only* in the Year 1 unit *The Overland Trail*. The notation will also be introduced in the Year 4 unit *The World of Functions*.

The calculator has this as a built-in function. The greatest integer function can be found by pressing MATH, highlighting **NUM**, and selecting **5:int**. Try each of the examples illustrated in this screen.

```
int(3.14)
             3
int(5)
             5
int(-6.3)
            -7
■
```

Slippery Slopes: If students have discovered the calculator's ability to determine the numerical derivative, they may be able to complete the derivative columns for this activity rather quickly. As noted previously, there are advantages and disadvantages to having students do this.

If students do suggest a possible equation that expresses the derivative in terms of either y or x, encourage them to use their calculators to verify that the rule works. For example, students may enter the function into the Y= editor and then look at the table. The table should match approximately the values for derivatives they calculated.

During discussions of *Slippery Slopes, How Does It Grow?,* and the "proportionality property" of exponential functions, it may be useful to quickly refer to the three-column tables students developed. This is not a skill that students should necessarily learn, but it may be helpful to you when facilitating these discussions.

If you enter $2x$ for **Y₁** and $0.693 \cdot 2^x$ for **Y₂** and then press 2ND [TABLE], the calculator will display the three-column table shown here.

X	Y₁	Y₂
0	1	.693
1	2.09	1.386
2	4.3681	2.772
3	9.1293	5.544
4	19.08	11.088
5	39.878	22.176
6	83.345	44.352

X=0

The Forgotten Account: Some students may have solved Question 3 from this activity by graphing and tracing the function defined by the expression $50 \cdot 1.045^x$.

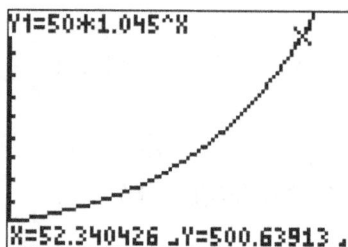

California and Exponents: The exponential function through the points (0, 92,600) and (10, 380,000) can be found using the regression features of the calculator. However, students should challenge themselves to find an exponential function algebraically in this activity.

To find the base for the exponential equation in Question 1, students must solve the equation $b^{10} = 4.1$ (or something equivalent). Suggestions are offered in the *Teacher's Guide,* but some students may describe something like "taking the square root, but the tenth instead." To find the 10th root of 4.1, enter **10** in the home screen. Then press $\boxed{\text{MATH}}$, highlight **5:x√**. Finish by entering **4.1** and pressing $\boxed{\text{ENTER}}$.

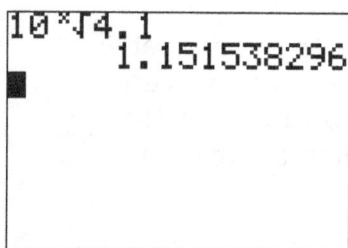

You can perform the equivalent operation by entering **4.1^(1/10)** and pressing $\boxed{\text{ENTER}}$.

Students may solve Questions 2 and 3 by graphing, zooming in, and tracing the graph of $y = 92{,}600 \cdot x^{10}$.

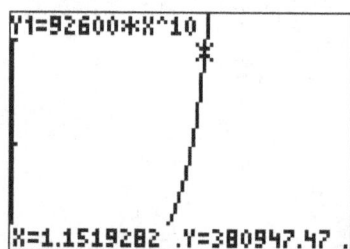

Comparing Derivatives: Some students might solve pieces or all of Part I using the numerical derivative feature of their graphing calculators. Encourage these students to consider the graphical perspective as well. Likely, they will have to for Part II.

If students are determined to find an approximate rule for the function graphed in Part II, encourage them to pursue this using the calculator's regression capability or by extending the matrix algebra from *Fitting Quadratics* in *Meadows or Malls?,* using some set of points on the graph (such as the five intercepts).

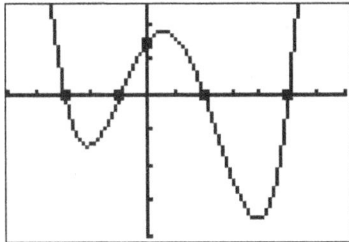

If students are able to generate a function that is similar to the graph from Part II, they can then have the calculator graph the numerical derivatives for this function, as shown here. (See the Calculator Note "Derivative at a Point" for help.) If any student investigates this, it is an exciting extension and a mathematical "wow!" for classmates.

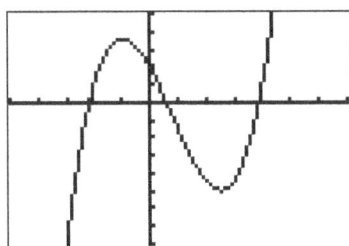

The Limit of Their Generosity: Students can investigate what happens as they use the calculator to compound more and more frequently. Enter the expression $\left(1 + \dfrac{1}{x}\right)^{x}$ into **Y₁** and investigate the graph or a table of values.

When entering this expression, be careful to follow the calculator's order of operations: enter it as **(1+1/X)^X**. When students trace, they will recognize the y-coordinate as the special number e.

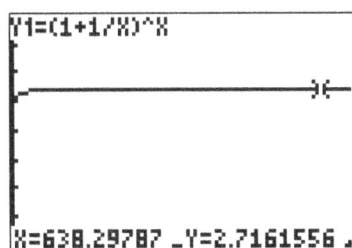

(If students have developed an equivalent expression, they should be encouraged to investigate it for large input values.)

To get e directly on the calculator, return to the home screen. Press 2ND [e^x] and then enter 1 (enclosing it in parentheses).

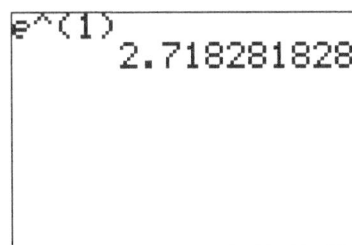

To determine the approximate value of ln(2) on the calculator, enter the expression exactly as shown here by pressing $\boxed{\text{LN}}$ $\boxed{2}$ $\boxed{)}$ $\boxed{\text{ENTER}}$.

```
ln(2)
     .6931471806
```

Tweaking the Function: This activity does not have to be done with the calculator's graphing capabilities, but students will likely identify the calculator as the appropriate tool to help them examine what their tweaking does. You might ask students, "How can you quickly compare graphs of two or three different functions?" Encourage students to graph two (or more) functions at a time in order to compare their differences.

Return to "A Crowded Place": This activity can be done using the graphing calculator. Students should plot all the data pairs and then test and compare the functions they create.

You might give students who are not yet comfortable with plotting points on the calculator the Calculator Note "Plotting Points," found in *Calculator Basics* in the Year 3 general resources.

Some students may have explored exponential regression on the calculator or asked about the how to determine a "best fit." They will do more work with best-fit curves and regression on the graphing calculator in the Year 4 unit *The World of Functions*. Students need not master calculator methods for determining best-fit lines during this unit.

"Small World, Isn't It?" Portflio: Students may have done a great deal of work on their graphing calculator during this unit and may want to include printouts of calculator screens in their portfolios. Students can use TI Connect™ software and connecting cable to connect the calculator to the computer to print screen captures. See the Calculator Note "Linking the Calculator to a Computer," found in *Calculator Basics* in the Year 3 general resources, for more information.

Assessments: Students should have access to graphing calculators during the in-class assessment.

Question 2a on the *In-Class Assessment* can be solved using the **nDeriv** command on the graphing calculator. We recommend that you have students show how to find the given derivative without the **nDeriv** feature.

Supplemental Activity—Slope and Slant: This activity is an interesting extension problem that relates right-triangle trigonometry to the algebraic concept of slope. However, students may see no connection if their calculator is in radian mode! Press $\boxed{\text{MODE}}$ and highlight **Degree**.

Supplemental Activity—Speedy's Speed by Algebra: In this activity, if students are able to create an algebraic expression in terms of h for Speedy's average speed during the interval from 50 – *h* to 50 (see Question 2), they may recognize that the same technique can be used to calculate Speedy's average speed for other intervals of length *h* seconds. Students may wish to explore this further using the function-graphing and table capabilities of their calculators. You might get them started by asking, "Could your ideas be used in conjunction with your calculator's ability to graph or make tables from rules?" Students may take this idea in many different directions.

Supplemental Activity—Proving the Tangent: After doing this activity, students may be curious about the graphing calculator's tangent features. Encourage them to use the calculator's guidebook to research this topic.

For your information, the following steps will yield the equation of a line tangent to a point on a curve. First, enter a function and reasonable window range. For example, if you want to find the derivative of $f(x) = 0.5x^2$ at the point (2, 2), you can use the window settings shown here.

```
WINDOW
 Xmin=-1
 Xmax=3
 Xscl=.2
 Ymin=-1
 Ymax=4
 Yscl=.5
 Xres=1
```

Then press GRAPH, which will take you to the graphing screen. Next, press 2ND [DRAW] and select **5:Tangent(**, which returns you to the graphing screen. Lastly, press 2 (for the value of *x*) and press ENTER. Displayed at the bottom of the screen is the equation of the line tangent to the curve at *x* = 2.

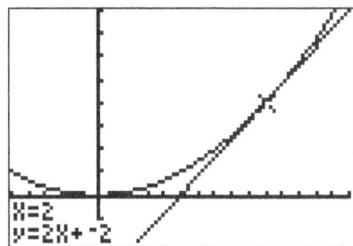

Use 2ND [DRAW] and **1:ClrDraw** to remove the tangent line.

Supplemental Activity—The Reality of Compounding: Students with an interest in investing money may wish to pursue this activity. Also, they may want to investigate the financial functions available on the graphing calculator. To find these functions, press APPS and select **Finance…**.

Derivative at a Point

The graphing calculator has a built-in command for estimating the numerical derivative of a function for a given input value. The method approximates the derivative by calculating the slope of the secant line through a point slightly below and a point slightly above the given input value. This is very similar to zooming in on a graph very closely and tracing two points near the input value.

Calculating a Numerical Derivative at a Point

Begin in the home screen. Press MATH, highlight **8:nDeriv(**, and press ENTER. Next, enter the function you are working with. Follow this with a comma, the variable, another comma, and then the value of the variable at which you want to evaluate the derivative. For Question 1 from *Speeds, Rates, and Derivatives*, you would enter the information as in the screen shown here. If the function $y = 400 - 16x^2$ is already in the Y= editor, then you can enter **nDeriv(Y₁,X,3)** instead of **nDeriv (400–16T²,T,3)** at the home screen.

```
nDeriv(400-16T²,
T,3)
              -96
■
```

Curve of Best Fit

You probably would like to make adjustments to your original estimate of the exponential function so that it models more closely the changes in the world's population over the past several centuries.

Plot the world population data set and set a reasonable window range. Next, enter your first guess as **Y1**.

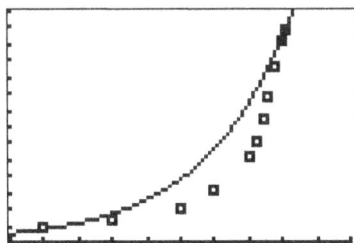

Improving Your Curve of Best Fit

Enter a new equation that is an improvement on your first try. If you like, leave your first guess in **Y1** and put your new guess in **Y2** so you can compare their graphs after pressing GRAPH. Continue improving on your best-fit exponential function until you are happy with it. But be patient, because this might require many tries.

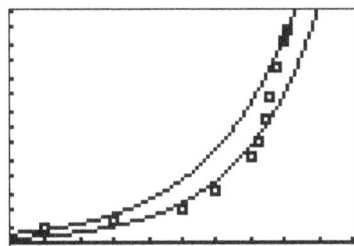

Using Your Curve of Best Fit to Make a Prediction

You can use your curve of best fit and the calculator's TRACE feature to predict what year the earth's population will reach 1.6×10^{15} people. You may need to adjust the size of your window!

Note: The TRACE feature moves the cursor along any graphed functions or scatter plots. If you find you are tracing the wrong function or plots, press the up- or down-arrow key to move the cursor to the correct function.

Another option is to press 2ND [TABLE]. You can adjust the options in the 2ND [TBLSET] menu to aid your search.

www.ingramcontent.com/pod-product-compliance
Lightning Source LLC
Chambersburg PA
CBHW051345200326
41521CB00014B/2481